DYNAMICS OF HERITAGE MANAGEMENT AND TOURISM DEVELOPMENT

Trends, Practices, and Approaches

Advances in Hospitality and Tourism

DYNAMICS OF HERITAGE MANAGEMENT AND TOURISM DEVELOPMENT

Trends, Practices, and Approaches

Surabhi Srivastava, PhD
Ruchika Kulshrestha, PhD
Editors

AAP | APPLE ACADEMIC PRESS

First edition published 2025

Apple Academic Press Inc.
1265 Goldenrod Circle, NE,
Palm Bay, FL 32905 USA

760 Laurentian Drive, Unit 19,
Burlington, ON L7N 0A4, CANADA

CRC Press
2385 NW Executive Center Drive,
Suite 320, Boca Raton FL 33431

4 Park Square, Milton Park,
Abingdon, Oxon, OX14 4RN UK

Library and Archives Canada Cataloguing in Publication

CIP data on file with Canada Library and Archives

Library of Congress Cataloging-in-Publication Data

CIP data on file with US Library of Congress

ISBN: 978-1-77491-804-3 (hbk)
ISBN: 978-1-77491-805-0 (pbk)
ISBN: 978-1-00359-226-6 (ebk)
DOI: 10.1201/9781003592266

ABOUT THE BOOK SERIES: ADVANCES IN HOSPITALITY AND TOURISM

Editor-in-Chief

Mahmood A. Khan, PhD
Professor, Department of Hospitality and Tourism Management,
Pamplin College of Business,
Virginia Polytechnic Institute and State University,
Falls Church, Virginia
mail: mahmood@vt.edu

This series reports on research developments and advances in the rapidly growing area of hospitality and tourism. Each publication in this series presents state-of-the-art information on specialized topics of current interest. Experts in respective field author/edit publications in the series. These one-of-a-kind publications are valuable resources for academia as well as for professionals in the industrial sector.

BOOKS IN THE SERIES

Brand Co-Creation Tourism Research: Contemporary Issues and Challenges
Editor: Raouf Ahmad Rather, PhD

Capacity Building Through Heritage Tourism: An International Perspective
Surabhi Srivastava, PhD

Chinese Outbound Tourism 2.0
Editor: Xiang (Robert) Li, PhD

Dynamics of Heritage Management and Tourism Development: Trends, Practices, and Approaches
Editors: Surabhi Srivastava, PhD, and Ruchika Kulshrestha, PhD

Dynamics of the Tourism Industry: Post-Pandemic and Post-Disaster Perspectives and Strategies
Editors: Gül Erkol Bayram, PhD, and Anukrati Sharma, PhD

Ecotourism and Sustainable Tourism: From Principle to Practice in the New Millennium
Editors: Kelly S. Bricker, PhD, and Jacqueline Kariithi, PhD

Event Tourism in Asian Countries: Challenges and Prospects
Editors: Shruti Arora, PhD, and Anukrati Sharma, PhD

Evolving Paradigms in Tourism and Hospitality in Developing Countries: A Case Study of India
Editor: Bindi Varghese, PhD

Food Safety: Researching the Hazard in Hazardous Foods
Editors: Barbara Almanza, PhD, RD, and Richard Ghiselli, PhD

Food Safety: Researching the Hazard in Hazardous Foods
Editors: Barbara Almanza, PhD, RD, and Richard Ghiselli, PhD

Himalayan Mountain Tourism: Toward Sustainable Adventure Tourism
Editors: Satish Chandra Bagri, PhD, and Junaid K. C., MBA

Hospitality Marketing and Consumer Behavior: Creating Memorable Experiences
Editor: Vinnie Jauhari, PhD

Labor in the Tourism and Hospitality Industry: Skills, Ethics, Issues, and Rights
Abdallah M. Elshaer, PhD, and Asmaa M. Marzouk, PhD

Management Science in Hospitality and Tourism: Theory, Practice, and Applications
Editors: Muzaffer Uysal, PhD, Zvi Schwartz, PhD, and Ercan Sirakaya-Turk, PhD

Managing Sustainability in the Hospitality and Tourism Industry: Paradigms and Directions for the Future
Editor: Vinnie Jauhari, PhD

Medical Tourism and Wellness: Hospitality Bridging Healthcare (H2H)©
Editor: Frederick J. DeMicco, PhD, RD

Medical Travel Brand Management: Success Strategies for Hospitality Bridging Healthcare (H2H)
Editors: Frederick J. DeMicco, PhD, RD, and Ali A. Poorani, PhD

Medical, Dental, and Wellness Tourism: A Post-Pandemic Perspective
Mary Schreiber Swenson, PhD, and Amit Bansal

Post-COVID Tourism and Hospitality Dynamics: Recovery, Revival, and Re-start
Editors: Umendra Narayan Shukla, PhD, and
Sharad Kumar Kulshreshtha, PhD

Poverty Alleviation through Tourism Development: A Comprehensive and Integrated Approach
Robertico Croes, PhD, and Manuel Rivera, PhD

Purpose-Driven Tourism: The Intersection of Cultural Cuisine, Gender Equality, and Sustainable Community Building
Editors: Andrea Edurne Jimenez Ruiz, PhD, and Shivam Bhartiya, PhD

Small Island and Small Destination Tourism: Overcoming the Smallness Barrier for Economic Growth and Tourism Competitiveness
Robertico Croes, PhD

Strategic Management for the Hospitality and Tourism Industry: Developing a Competitive Advantage
Vincent Sabourin, PhD

Strategic Winery Tourism and Management: Building Competitive Winery Tourism and Winery Management Strategy
Editor: Kyuho Lee, PhD

Sustainability, Social Responsibility, and Innovations in Tourism and Hospitality
Editor: H. G. Parsa, PhD; Consulting Editor: Vijaya "Vi" Narapareddy,
PhD; Associate Editors: SooCheong (Shawn) Jang, PhD,
Marival Segarra-Oña, PhD, and Rachel J. C. Chen, PhD, CHE

Sustainable Tourism Development: Futuristic Approaches
Editor: Anukrati Sharma, PhD

Sustainable Viticulture: The Vines and Wines of Burgundy
Claude Chapuis

The Hospitality and Tourism Industry in ASEAN and East Asian Destinations: New Growth, Trends, and Developments
Editors: Jinlin Zhao, PhD, Lianping Ren, D.HTM, and Xiangping Li, PhD

The Hospitality and Tourism Industry in China: New Growth, Trends, and Developments
Editor: Jinlin Zhao, PhD

The Indian Hospitality Industry: Dynamics and Future Trends
Editors: Sandeep Munjal and Sudhanshu Bhushan, PhD

Tourism in Central Asia: Cultural Potential and Challenges
Editors: Kemal Kantarci, PhD, Muzaffer Uysal, PhD, and Vincent Magnini, PhD

Tourism in Turkey: A Comprehensive Overview and Analysis for Sustainable Alternative Tourism
Editor: Ahmet Salih İKİZ

Tourist Behavior: Past, Present, and Future
Editors: Narendra Kumar, PhD, Bruno Barbosa Sousa, PhD, and Swati Sharma, PhD

Wilderness of Wildlife Tourism
Editor: Johra Kayeser Fatima, PhD

Women and Travel: Historical and Contemporary Perspectives
Editors: Catheryn Khoo-Lattimore, PhD, and Erica Wilson, PhD

ABOUT THE EDITORS

Surabhi Srivastava, PhD
Associate Professor, Om Kothari Institute of Research and Management, Kota, Rajasthan, India

Surabhi Srivastava, PhD, is a scrupulous and analytical researcher with 18 years of teaching experience in heritage tourism marketing and management. She is an Associate Professor at the Om Kothari Institute of Research and Management, Kota, Rajasthan, India. She has organized workshops, seminars, and presentations, educating students, peers, and superiors on exploring new trends in tourism development. She has contributed as the author of four books (national publications) and editor of two books (international publications) on heritage tourism, dedicated toward research findings, gaps, and limitations of heritage tourism academia. She has been awarded first prizes for her research and also received the Global Women Inspiration Award (2022) for exploring tribal tourism, nature tourism, and agri-tourism for community development. She holds an MBA (tourism) and a PhD (Tourism) from the University of Lucknow, India, and a PDF (Tourism) from the University of Kota (UGC-sponsored), India.

Ruchika Kulshrestha, PhD
Assistant Professor, Institute of Business Management, GLA University Mathura, India

Ruchika Kulshrestha, PhD, is working as an Assistant Professor at the Institute of Business Management at GLA University Mathura, India. Her research interest pertains to heritage tourism, destination management, tourism marketing, and tourist behavior. She has published her research work in reputed international Scopus-indexed journals and also presented papers at various international conferences. Her articles are published in the Economic Times, Travel Daily News, and other magazines. She had the opportunity to be a session chair at IIM Kashipur and DSMS Durgapur, among others.

She has received various awards for her contribution to tourism education. She is actively involved in organizing faculty development programs. She has earned her PhD (Tourism) from Jiwaji University, MBA (Tourism) from IIPS Indore, and is qualified in UGC NET (Tourism Management).

CONTENTS

CONTRIBUTORS

Fernando Almeida-García
Associate Professor Faculty of Tourism, University of Malaga, Spain

José Álvarez-García
Full Professor and Researcher, Accounting and Financial Economy, Department of the Faculty of Business, Finance and Tourism, University of Extremadura, Spain

Anubha
Research Scholar, Department of Tourism and Hospitality Management, Jamia Millia Islamia (A Central University), New Delhi, India

Shankha Pratim Bhattacharya
Associate Professor, Architecture and Regional Planning, IIT, Kharagpur, West Bengal, India

Sarang S. Dhote
Assistant Professor, Department of Chemistry, S.S.E.S. Amravati's Science College, Nagpur, Maharashtra, India

Tsegaye Ebabey
Department of Anthropology, Hawassa University, Ethiopia

Rosse Marie Esparza-Huamanchumo
Research Group in Gastronomy and Culinary Art, Faculty of Hotel Administration, Tourism and Gastronomy, Universidad San Ignacio de Loyola, Lima, Peru; Doctoral Program in Sustainable Territorial Development (R015), International Doctoral School, University of Extremadura, Badajoz, Spain

Martha Beatriz Flores-Romero
Faculty of Accounting and Management, Saint Nicholas and Hidalgo Michoacán State University (UMSNH), Morelia, Mexico

Isabel Vaz de Freitas
Professor and Head of the Department of Tourism, Heritage, and Culture, Universidade Portucalense, Portugal

Sarah Hussain
Associate Professor, Department of Tourism and Hospitality Management, Jamia Millia Islamia (A Central University), New Delhi, India

Donaji Jiménez-Islas
Renewable Energy Division, TecNM: Higher Technological Institute of Huichapan, Huichapan, Hidalgo, Mexico

Radmila Jovanović
Faculty of Tourism, University of Malaga, Spain

I. M. Karthikeyan
Assistant Professor, Department of Tourism and Hospitality Management, Annai Fathima College of Arts and Science, Thirumangalam, Madurai, Tamil Nadu, India

Sharad Kumar Kulshreshtha
Assistant Professor, Department of Tourism and Hotel Management, North-Eastern Hill University, Shillong, Meghalaya, India

Ruchika Kulshrestha
Assistant Professor, Institute of Business Management, GLA University, Mathura, Uttar Pradesh, India

Krishna Kumar
Research Scholar, Center of Excellence for Indian Knowledge Systems, IIT, Kharagpur, West Bengal, India

Yonas Legesse
Department of Anthropology, Hawassa University, Ethiopia

Rafael Cortés Macías
Professor of Human Geography at the University of Malaga, Spain

Pulak Mishra
Professor (Economics), Department of Humanities and Social Sciences, IIT, Kharagpur, West Bengal, India

Priyakrushna Mohanty
Assistant Professor, Department of Business and Management, Christ University, Bangalore, Karnataka, India

Nam Trung Nguyen
Researcher at Vietnam Tourism Research, Vietnam

Nga Viet Nguyen
Faculty of Political Theory and General Knowledge, Ho Chi Minh City University of Culture, Vietnam

Miriam Edith Pérez-Romero
Business Management Division, TecNM: Higher Technological Institute of Huichapan, Huichapan, Hidalgo, Mexico; Doctoral Program in Sustainable Territorial Development (R015), International Doctoral School, University of Extremadura, Badajoz, Spain

José David Albarrán Periáñez
Independent Researcher and Consultant, PhD in Geography, Complutense University of Madrid, Spain

Tuyen Dai Quang
Faculty of Tourism and Vietnamese Studies, Nguyen Tat Thanh University, Vietnam

Makhabbat Ramazanova
Assistant Professor and Coordinator, Tourism and Hospitality Universidade Portucalense, Portugal

María de la Cruz Del Rio-Rama
Business Management and Marketing Department, Faculty of Business Sciences and Tourism, University of Vigo, Ourense, Spain

Pablo César Romo-Román
Faculty of Business Sciences of the Universidad San Ignacio de Loyola, Lima, Perú

Javed Salim
Research Scholar, Department of Tourism and Hospitality Management, Jamia Millia Islamia (A Central University), New Delhi, India

Bao Ha Son
Researcher at Vietnam Tourism Research, Vietnam

Surabhi Srivastava
Associate Professor, Om Kothari Institute of Research and Management, Kota, Rajasthan, India

Jonel Subić
Principal Research Fellow, Institute of Agricultural Economics, Belgrade, Serbia

Iadonlang Tynsong
Faculty, Department of Tourism and Travel Management, Martin Luther Christian University, Shillong, Meghalaya, India

Thanh Van Vo
Faculty of Tourism, Van Lang University, Ho Chi Minh City, Vietnam

Predrag Vuković
Senior Research Associate, Institute of Agricultural Economics, Belgrade, Serbia

Nusrat Yasmeen
Assistant Professor, Department of Tourism and Hospitality Management, Jamia Millia Islamia (A Central University), New Delhi, India

ACKNOWLEDGMENTS

Heritage management is defined as the integration of both the product and the process, and when we amalgamate it with sustainable tourism, it becomes of utmost significance for the social and cultural landscape. Keeping all this in mind, I was determined to prepare a manuscript that incorporates the global perspectives of all dimensions of heritage management. This journey takes shape with the pillar, my colleague, a heritage tourism scientific researcher, Dr. Ruchika Kulshreshtha. She has a high level of determination and enthusiasm to work on the cultural heritage and community and develop strategies for the conservation of built and living heritage. With her joining hands, this manuscript has its present shape.

We both are greatly indebted as editors of this book to the original thinkers and contributors to the subject. We acknowledge with gratitude the contributions of the writers, researchers, and thinkers who elucidated the theoretical and practical aspects of Heritage Management, which enabled us to compile this manuscript. We deeply bestow our gratitude to the contributors for sharing their pearls of wisdom and providing insight and expertise that greatly assisted the research.

With the grace of God and the blessings of our elders and our senior professors, we have shaped this manuscript into a book. Their continuous inspiration, supervision, and vast knowledge facilitated us to prepare this present volume.

I owe Apple Academic Press a great debt of appreciation for putting this manuscript in its current state. In particular, I am grateful to Dr. Mahmood A. Khan, the series editor, without whose oversight, direction, and constant support, this book would not have been published.

Last but not least, we owe a debt of gratitude to our parents, husband, and loving children for their unwavering support and perseverance shown to us throughout the protracted period of our study. Throughout our work, their patience and optimism have always been motivating.

I feel very delighted to express my gratitude toward my friends, colleagues, and seniors for creating this book.

—Surabhi Srivastava, PhD
Ruchika Kulshrestha, PhD

PREFACE

The pandemic period taught us about the value of our heritage. Today, researchers are focusing on glocalization, and it all begins with the management of heritage resources. Presently, managing and conserving heritage resources and presenting them as a tourism product for sustainable development are major global issues for conservators, environmentalists, and tourism developers. Keeping all this in mind, the editors of this book planned to develop a platform where global researchers, practitioners, and developers can join and share their perspectives regarding conserving and managing heritage for tourism development. The effort is to bring together all aspects of heritage and tourism at a global level with various case studies and perspectives. The focal intent of the book is to present an assortment of dynamics for identifying, managing, and conserving heritage tourism. Heritage tourism can be an interesting way of learning history and culture, even virtually, as we have seen during the COVID-19 pandemic. Managing heritage tourism will raise global awareness as well as profound regard for heritage sites. This book also emphasizes the responsibility for regulating the dynamics of historic tourism, which many believe is the duty of the government rather than individuals.

What is heritage? While answering this question, the mind wonders whether all past is heritage or documented history is heritage. Who wants to conserve the heritage? What are the assets that should be passed down through generations? Much brainstorming and literature resulted in the conclusion that not everything is heritage, but anything could become heritage. That means we are expected to decide to craft it as our heritage. The literary meaning of heritage is inherited or legacy. So, whatever we inherited is our heritage? The answer is No. There is always the question of how to choose an object or tradition from the vast historical past to accept as heritage. Heritage Studies and Heritage Management are complex but growing fields in scholastic disciplines. This essentially says that heritage is the end result of an intended artistic endeavor. The elementary criteria for conserving or protecting the past are not all that we have received from past generations, but just that significant portion that we decide to own and accept responsibility for.

Heritage and History

The words "past," "history," and "heritage" are distinct. The past encompasses everything that has ever happened, and history has been considered to be an interpretation of some aspect of the past by contemporary historians. In the context of heritage, it is crucial to contemplate how past events are interpreted and expressed in the present. The expression "heritage" incorporates human cultural traits, which might be tangible or intangible, mobile or stationary, contemporary or antique, and individually, collectively, or simply not owned. The dynamic culture of communities, the elements of their essential character, and their diversity and uniqueness are represented through intangible forms that are highly relevant to various civilizations. People's lives, thoughts, sentiments, aspirations, and recollections are shaped by their heritage, which also teaches us about historical cultures and communities. It is also an efficient approach for extracting lessons from past events.

Heritage is a domain of active public contemplation, debate, and discussion. It is both concrete and abstract. It can serve as a pillar of regional and urban design, a stage for civic acknowledgment, a forum for intercultural communication, and the foundation for local economic growth. Nostalgia has historically been viewed as a troublesome, unreliable, and unproductive emotion. It has also been ignored as a study topic and recognized as an unreliable framework for heritage buildings.

The phrase "cultural legacy," which is sometimes used for cultural heritage, can be defined as customs, rituals, traditions, values, performing arts, crafts, historical values, buildings, sculptures, and wall paintings, as well as historical, scientific, anthropological, and artistic values. Allowing interested individuals to physically interact with all invaluable assets like sites, events, or artistic performances is one approach to making cultural heritage accessible. This brings tourists to the area around the heritage components and to the site itself. Heritage has a major influence on how people live, feel, and remember. It can give us insight into cultures and peoples of the past, what their lives were like, the obstacles they overcame, and how they accomplished them. Heritage can be a powerful tool for gaining insight from the past. It is essential that the lessons learned be utilized to comprehend and address current and upcoming global issues, such as climate change, migration, discord, and decolonization.

When humans refine and redefine something that is naturally occurring, it progressively assumes new dimensions in the lives of the individuals and even more so at the interface between the individuals and society. A cultural heritage is characterized by its cultural values, identity, relationships, and power to evoke memories in addition to its historical importance and aesthetic attributes.

Assessment, preservation, and management of heritage landscapes as in public benefit are the focus of the evolving field of heritage management. The structure and application amalgamation are recognized as heritage management that enables society to benefit from resources that were developed and passed down from one generation to the next, as well as from those that are currently being produced and bequeathed for future generations. It provides information on people, politics, religion, architecture, and technology. The involvement of residents opens up opportunities for the tourism industry as well.

Contemporary heritage management is more about communities and their ideologies than physical structures, artifacts, and sites. Expertly espoused intrinsic heritage value systems—those related to history or the arts—that traditionally profoundly influenced heritage management have given way to values that serve more functional reasons, such as the economic model, the social dimension of the community, and the environment. The promotion of heritage serves as a vehicle for the preservation and protection of natural, historic, and cultural resources as well as for enabling nearby communities' substantial economic avenues. Heritage resources and communities could reap benefits from heritage tourism.

The present volume is an anthology of conceptual analysis and its implications with a number of case studies, research analysis on heritage management with community involvement for tourism development. The main objective of this book is to explore the various dynamics of identifying, managing, and conserving heritage resources. Heritage tourism can be an interesting way of learning and conserving history and culture, even virtually, as we have seen during the COVID-19 pandemic. Managing heritage tourism has raised global awareness as well as profound regard for heritage sites. The book also emphasizes the responsibility for regulating the dynamics of historic tourism, which many believe is the duty of the government rather than individuals.

This book offers a thorough assessment of the literature on heritage management to highlight the significance of conserving and promoting tangible and intangible cultural assets for sustainable tourism development. It also aims to provide guidance for destinations experiencing problems with heritage management or for developing tourism destinations to help them avoid similar issues.

This edited volume is divided into three sections that incorporate and elucidate all components of heritage management. Part I includes research chapters discussing the changing dynamics of Heritage Management and Sustainable Tourism.

The first chapter of the book is the theme of this book and encompasses the conceptual note of changing dynamics of Heritage Management through sustainability, which is authored by Dr. Ruchika Kulshreshtha and Dr. Surabhi Srivastava, the editors of this volume. This chapter discusses the need for opportunities and challenges for heritage preservation for sustainable development. The social and cultural aspects related to heritage are explained in these changing dynamics for future generations, showing how sustainability increases the heritage assets' usability to ensure their continued usability. The conservation and resilience of heritage are mentioned along with the development aspects to balance cultural and value-based heritage.

A senior group of four professors from Europe (Peru and Spain) worked on birdwatching as a tourist activity. The second chapter incorporates the investigation at the Arena Blanca Ecological Reserve, a site specialized in providing bird-watching services located in Alto Mayo, San Martín, Peru. The objective is to analyze the implementation of continuous heritage interpretation processes that facilitate birdwatching, as well as the management of the tourist site. This chapter concludes with a mixed methodological basis that allowed the interpretation to be used as a management tool for the tourist site and evaluated the effectiveness and satisfaction of the operators, tourists, guides, and/or visitors in the visiting area.

The third chapter explores the challenges of preserving and promoting architectural heritage values for tourism development in urban areas, with a focus on Ho Chi Minh City. It highlights the need for the state management agency to play an intermediary role and bring tourism in the right direction. The balance between preserving architectural heritage and developing tourism in urban areas is a long-standing challenge not only in Vietnam but also in other countries worldwide. The chapter explores the possibilities of new technologies like virtual reality and augmented reality in promoting architectural heritage for tourism growth. In addition, detailed research is required to evaluate the efficacy of present conservation initiatives and propose areas for improvement.

The fourth chapter talks about the adoption and acceptance of the concept of digitized monuments that are still surrounded by many unanswered questions. The chapter reveals the opinions of the tourists visiting the various built heritage sites and understand their experiences with digital heritage and its influence on the tourist experience while visiting a built heritage monument. The authors of this chapter are a group of young, energetic researchers who present the qualitative survey results that demonstrate how heritage consumption is laden with experiential value. Tourists believe that the experiences they currently have or have had in the past while visiting heritage monuments have an emotional and sensory resonance. They came to the conclusion that there is a lot of potential for producing immersive experiences by exploiting technology and digital interfaces effectively.

The fifth chapter acknowledges how romantic iconography continues to permeate and shape how people nowadays perceive Andalusia. The chapter came to a conclusion with the awareness that the tourism industry's advertising materials had a profound impact on local identity values, initiating a process that might transform how the people who actually live there view the place. Whether they are cognizant of it or not, inhabitants are burdened by a persistent stereotype in today's culture that eventually incorporates and seeps deeply into their own particular identities.

The sixth chapter emphasized how culture is a reflection of how history and tradition have changed over time. In addition to protecting and maintaining sacred sites, heritage resource management also recognizes that these sites have evolved and endured as a result of ongoing pilgrimages and rituals that trace back to early human interactions with the earth and its mystical power. The Sustainable Management of Pilgrimage and Heritage Tourism to Varanasi, an Indian Sacred City, serves as the major framework for this chapter. It views religion and religious literature from the standpoint of its utility for survival as an individual or group. The current study focuses on tracking the evolution of literary allusions used to promote pilgrimage-focused religious travel to Varanasi.

In the seventh chapter, a framework for heritage tourism planning was proposed incorporating NHA (National Heritage Areas) planning guidelines and statistical surveying techniques that can be used to more clearly illuminate legacy planning and the board in NHAs and other legacy regions around the world. Future studies on heritage tourism as a subset of a larger tourism population will build on the foundation laid out in this chapter. A relatively innovative and creative strategy for resource conservation in this sector is the combination of natural and cultural conservation with explanation. The results of this study show that historical tourism is a unique travel phenomenon compared to the mainstream tourist industry, in contrast to other tourism niches that have been thoroughly researched in the past.

Part II focused on the chapters that give the reflection of branding of heritage resources for sustainable development.

The researchers of the eighth chapter discuss Mexico's magical towns, which are popular tourist attractions and have a rich natural, cultural, and historical legacy. The chapter discussed the potential for cultural tourism present in Mexican villages with the classification of Magical villages. In order to more accurately assess the tourist potential of these locations, it is also important to include the opinions of the many players involved. Additionally, it would be beneficial to provide satisfying and distinctive experiences in these unique locations in order to forge an emotional connection with the traveler and positively impact others.

The professors of the ninth chapter offer an introduction to the idea of military tourism through a review of the literature because the growth of this kind of tourism has had a considerable impact on regional economies, particularly in places with a long military tradition. It also investigated whether or not military tourism attractions can promote world peace. The chapter demonstrates through in-depth research how, using military heritage tourism as an example, tourism can actively contribute to increasing responsibility for global economic growth (goal 8), promoting the reduction of inequalities between places and communities (goal 10), developing sustainable cities and communities (goal 11), and especially can make a difference in peace and justice (goal 16).

The tenth chapter introduced an app that provides details about historical statues and monuments. This is the first Central Museum in India to deploy such an app. This is presently available in three languages: Marathi, Hindi, and English. Users can select any of these languages, and the software will work in that language. Without an internet connection, this app still works. In India, this app has copyright protection as of 2022. After scanning the QR code or selecting the number associated with each sculpture, historical artifact, or distinctive object using this app, the information is accurate on the objects themselves. The chapter described how every museum visitor can be engaged with a smartphone application that is based on the current exhibition.

Part III encompassed the social and cultural aspects of heritage management at tourist destinations. The professors and thinkers in the eleventh chapter describe the new regionalization of Serbia's viticulture that took place at the start of the twenty-first century. It is clear from this viticultural regionalization that every area that produces wine contains smaller or bigger urban communities within its boundaries. The paper highlighted Serbia's wine-growing region's cultural heritage. The findings demonstrate that the majority of monasteries, which fall under the category of cultural properties of exceptional importance, are located in areas with the least anthropogenic influence. Fewer monuments fall within this group, whereas churches and museums are scarce. Contrarily, the region with the greatest human influence, where the natural landscape has been most significantly altered by human activity, is home to the greatest number of cultural heritages in the category of other cultural assets.

The objective of the thinkers and researchers in the twelfth chapter is to redirect the present trend towards a collective-based legacy strategy by analyzing the rhetoric and subsequent effects of this idea on historical and cultural legacies of Ethiopia. In order to use heritage for the preservation of Ethiopians' threatened coexistence and community development practices, it is important to emphasize the heritagization of cultural heritage, its interpretation within a context of collective memory, as well as the appreciation and utilization of the diverse forms of cultures. With an emphasis on the ethnic-based perspective of cultural legacies and its various detrimental effects for Ethiopian unity, this chapter explored the tendency of heritagization and interpretation in Ethiopia. However, Ethiopia's past offers more opportunities than challenges to support its constructive role in society today.

The thirteenth chapter of this book analyzes and emphasizes the value of fostering solid collaboration between the Cham community of Vietnam and the local government in order to sustainably promote tourism at the living-sacred sites of the Cham community. The suggested management framework emphasizes the importance of involving local participation in decision-making processes to foster culturally sensitive and sustainable tourism growth. The study's conclusions add to the ongoing discussion over heritage preservation and tourism development, particularly in communities that are home to ethnic minorities like the Cham people. This chapter highlights the challenges that communities of color confront in terms of preserving their cultural heritage and developing tourism. The study has improved our understanding of power dynamics among tourism sector players, particularly in locations with ethnic or indigenous history, and how they can affect the effectiveness of collaborations.

The primary goals of the professor authors of the fourteenth chapter are to identify the challenges faced by the Khasi community (India) in maintaining their intangible cultural heritage in the face of advancements and globalization, to investigate the potential of event tourism in promoting Nongkrem Festival as a tourist destination while ensuring the preservation of its cultural values and traditions, and to develop guidelines and recommendations for promoting the Nongkrem Festival. Deductive reasoning was employed to arrive at the study's conclusions, which were obtained utilizing a qualitative technique. The chapter concluded with this strategy can ensure that the event is preserved, and that cultural practices and traditions are upheld and promote responsible tourism.

The book provides in-depth knowledge by covering dynamics from traditions to digitization. The chapters focus on and approach bird watching, tirtha darshan, conservation, military tourism, etc. What we acquire from previous generations, both material and immaterial, constitutes our cultural heritage. The roots of these trees go deep, and their lifespans are impressive. We cannot make sense of the world without them, and yet they endure practically unchanged from one generation to the next, long after the economic and social demographic conditions that gave rise to them have faded.

Sharing economies, in which people are the most vibrant and sensible resource, require safety and security, protection, and precaution during the visit at heritage sites and other activities with various stakeholders. With the advent of the new age of heritage and an increase in advanced information and communication technology, tourists are providing economies to the various stakeholders of the destination.

PART I

CHANGING DYNAMICS OF HERITAGE MANAGEMENT AND SUSTAINABLE TOURISM

CHAPTER 1

CHANGING DYNAMICS OF HERITAGE MANAGEMENT THROUGH SUSTAINABILITY

RUCHIKA KULSHRESTHA[1] and SURABHI SRIVASTAVA[2]

[1]*Assistant Professor, Institute of Business Management, GLA University, Mathura, Uttar Pradesh, India*

[2]*Associate Professor, Om Kothari Institute of Research and Management, Kota, Rajasthan, India*

ABSTRACT

A transformative strategy that ensures the long-term viability and significance of preservation and conservation of heritage and its products can be implemented by integrating sustainable development principles into heritage management practices. In the wake of the pandemic, this study investigates the adaptation of heritage management systems in the context of sustainable development, with an emphasis on the need to strike a balance between the demands of modern society and conservation efforts. The study underscores the ability of innovative technology and strategies to enhance the branding and marketing of traditional products, while simultaneously encouraging sustainable practices. The future of heritage management and sustainable strategies for promoting community engagement in heritage preservation are the primary topics of interest. This strategy aims to promote economic development, enhance social well-being, and safeguard cultural heritage for future generations by incorporating sustainability into its management. The

Dynamics of Heritage Management and Tourism Development: Trends, Practices, and Approaches.
Surabhi Srivastava & Ruchika Kulshrestha (Eds.)

article concludes by offering a set of suggestions for policymakers, heritage experts, and stakeholders to work together to establish a sustainable heritage management paradigm.

1.1 INTRODUCTION

When it comes to adapting to the continuing processes of global transformation and guiding it toward a brighter future, heritage has a role that is not just substantial but also of the utmost importance to play. Archaeologists and heritage industry professionals shouldn't keep trying to think and behave as if the world is still molded by the same forces that promoted the growth of archaeology and the heritage industry when they are steering these processes. A new generation of heritage managers has to have it ingrained in their brains that they are responsible for overcoming chauvinistic and haughty attitudes, sectarian biases, and militant ethnic and religious groupings. They should encourage, through their heritage, the principles of equality, peace, and knowledge; sociality; respect for others and cultural variety; creativity; aesthetic appreciation; emotional fulfillment; and respect for the natural world. They should also be concerned with the negative aspects of the commercialization of legacy as well as the well-being of the communities in the area where the heritage is located. This necessitates the implementation of an objective approach to cultural heritage management that is in sync with the new international priorities and for heritage professionals to be more engaged in social, governmental, as well as international forums. In addition, this calls for heritage managers to be more proactive in the preservation of cultural heritage.

Over the past few decades, discussions about heritage and tourism have been increasingly commonplace. Their connection is always fraught with tension since ecologists and preservationists view heritage tourism as undermining their cause for financial gain (Aas & Ladkin, 2005). Tourism has the potential to boost economies in culturally rich areas, but it also poses risks to the local culture, especially intangible cultural assets (Yang et al., 2010). Cultural heritage management is the practice and vocation of managing cultural heritage. Cultural heritage can be marked as the trends and practices of cultural restoration, conservation, archaeology, museology, architecture, and history (Orr et al., 2021) (Figure 1.1).

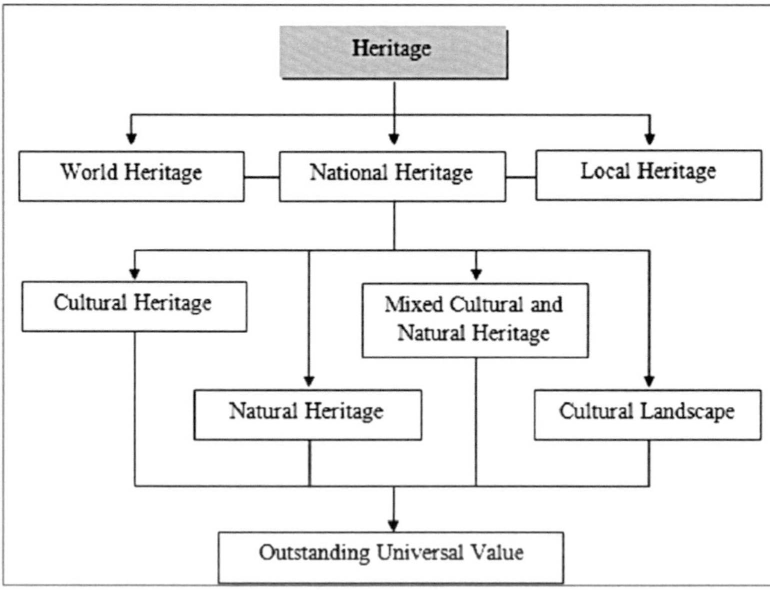

FIGURE 1.1 Classification of heritage from operational guidelines.

Source: Reprinted from Idris et al, 2016. https://creativecommons.org/licenses/by-sa/4.0/

According to Boniotti (2023), conserving cultural heritage can be considered a common good within the public sector purview, as well as involving the participation of private resources, which presents a good opportunity for public administration to benefit from a new funding channel. The chapter conveys the possible dynamics of heritage management related to sustainable conservation, approaches, and practices. Cultural heritage management also involves striking a balance between showcasing development practices to the tourism sector.

Value-based heritage conservation aims to preserve the cultural significance of tourist places by balancing scientific, historic, spiritual, and social values "held by past, present, and future generations" (Avrami et al., 2019). Therefore, it can be said that heritage management is a driving force for tourism and global development. In most cases, balancing mechanisms help in maintaining and protecting heritage as part of the economy. Heritage management is crucial as cultural and traditional monuments decline and deteriorate over time. However, sustainable development must be integrated into policies and practices related to heritage to achieve a comprehensive and successful incorporation of heritage into sustainable development goals and methods. Global heritage regulation regimes have made references to sustainable development since the mid-1990s (ICOMOS, 1996).

Therefore, it is significant to hold on to the values, beliefs, morals, and cultural practices of heritage to impart these to the next generation ensuring its core maintenance.

1.2 CHANGING DYNAMICS OF HERITAGE MANAGEMENT WITH TOURISM DEVELOPMENT

1.2.1 NEED FOR HERITAGE PRESERVATION AFTER THE PANDEMIC

The COVID-19 pandemic has had an influence on almost all sectors across different regions on a global level, along with impacting thousands of World Heritage properties. According to a survey conducted by UNESCO, it was observed that 90% of countries shut down the heritage sites, leading to large impacts on the local community due to a 40% reduction of permanent staff and a 53% reduction of temporary staff (UNESCO, 2021a). Hence, the negative influence of COVID on the local community and heritage contributes to an increased requirement for the preservation of heritage on a global level. The closing of World Heritage properties has led to the disruption of community life, socio-cultural impacts, reduced maintenance, an increase in illegal activities, and subsequent issues in the monitoring of conservation (UNESCO, 2021b). Therefore, the pandemic has led to increased challenges in implementing the World Heritage Convention and appropriate actions, thereby leading to an increased level of poverty.

The impact of COVID-19 on heritage has affected the benefits of local communities and tourism incomes, leading to a critical impact on the tourism industry. Therefore, the pandemic has led to significant uncertainties in the realignment of policies for tourism and stakeholders in the short term (UNESCO, 2021a). The pandemic has an influence on the suspension of cultural events that are organized on certain heritage sites, thereby affecting the revenue generated through them. Hence, the data from UNESCO reflects the subsequent impact of the pandemic on the negative influence on world heritage, which increases the need for the preservation of heritage for recovery through the losses faced by communities, the economy, and the tourism industry due to the shutting down of the heritage.

1.2.2 NEED FOR HERITAGE PRESERVATION FOR SUSTAINABLE DEVELOPMENT

Cultural heritages are a distinctive source of inspiration and play a major role in sustainable development. According to Miller-Rushing et al. (2021), cultural heritage has a critical role in keeping people connected with traditions, beliefs, religion, and community identities. Therefore, preserving world heritage is not only crucial for communities but also for nations as a whole. There is a correlation between World Heritage and economic development that not only supports communities but also leads to revenue generation for countries (UNESCO, 2021c). This information highlights the importance of heritage for sustainability, the economy, local communities, culture, and society. Therefore, there is an increased need for the preservation of heritage post-pandemic to ensure that it can provide substantial benefits to the economy and society on a global level. The 2030 Agenda adopted by the UN emphasizes the distinctive role of cultural heritage and cultural creativity as a driving force behind sustainable development (UNESCO, 2023a) (Figure 1.2).

FIGURE 1.2 Engaged bodies of sustainable heritage development.

Source: Reprinted from Sabou (2012) https://creativecommons.org/licenses/by/3.0/.

As per the above information, World Heritage has significant potential for developing and testing new approaches that could lead to sustainable development on a global level. Therefore, according to Foster (2020), world heritage acts as an anchor within the tourism industry on a global level, along with acting as a major contributor to the economy and sustainability. World Heritage is critical for economic development through the attraction of investments, thereby ensuring green and locally-based stable jobs, including tourism. Thus, well-maintained heritage acts as an indicator of natural as well as man-made mishaps, thereby increasing resilience within communities and promoting sustainable development (UNESCO, 2023a). Hence, these factors lead to an increased requirement for the conservation and preservation of world heritage to minimize risks of disasters, along with preserving values, history, cultural values, and economic values on a global level.

1.2.3 NEED FOR HERITAGE AND SUSTAINABILITY

Due to the fact that sustainability "is moving from a theoretical notion to a quantifiable state of changing human-ecological systems" (Mayer, 2008). At the moment, a significant amount of work is being put toward analyzing the effects of interdependencies in order to determine how well we are doing in terms of sustainable development (Nilsson et al., 2048). Indices and indicators are typically utilized in order to conduct extensive research into the results of assessments that focus on the causal linkages that exist between policies and goals at the local or municipal level. In a similar vein as in holistic management practices, there is a shortage of multidimensional systemic heritage processes and methods for the evaluation of synergies across wider urban sectors. Developing the cities is tough since their treasures are valuable, and the act of doing so may create vulnerabilities for natural, urban, and historic resources that should be maintained.

Comprehending one's own historical, cultural, societal, and traditional background is of paramount importance, and the field of heritage architecture assumes a pivotal role in facilitating this understanding. Elshater (2021) emphasized the significance of enhancing community awareness regarding the value of historic possessions and incorporating this aspect in forthcoming conservation research to prioritize local character over contemporary artificial ones and promote a sense of affiliation and individuality. According to Uzzell (1996), the preservation of cultural artifacts plays a critical role in fostering individuals' sense of self and community. Hence, it is imperative to consider this objective during the development and execution of conservation strategies. The maintenance of one's cultural and geographical identities is heavily reliant on the safeguarding of their heritage. To be sustainable, their growth must strike a balance between people's desires to improve their quality of life through urban and environmental improvements and their fears of losing touch with the natural and cultural landscapes that gave rise to them.

1.3 BRANDING AND MARKETING OF HERITAGE TOURISM PRODUCTS

1.3.1 MOST RECENT AND PROMISING PARTS OF HERITAGE MANAGEMENT

Comprehending one's own historical, cultural, societal, and traditional background is of paramount importance, and the field of heritage architecture assumes a pivotal role in facilitating this understanding. Elshater (2021) emphasized the significance of enhancing community awareness regarding the value of historic possessions and incorporating this aspect in forthcoming conservation research to prioritize local character over contemporary artificial ones and promote a sense of affiliation and individuality. According to Uzzell (1996), the preservation of cultural artifacts plays a critical role in fostering individuals' sense of self and community. Hence, it is imperative to consider this objective during the development and execution of conservation strategies. The maintenance of one's cultural and geographical identities is heavily reliant on the safeguarding of their heritage. To be sustainable, their growth must strike a balance between people's desires to improve their quality of life through urban and environmental improvements and their fears of losing touch with the natural and cultural landscapes that gave rise to them.

Therefore, providing rewards and recognition are critical promoting factors that contribute to driving an influence on nations for restoration and preservation of heritage property, thereby acting as a critical trend in heritage management in recent years. The awards give individuals a sense of ownership and pride in their heritage post-pandemic, hence it could have an influence on heritage management post-COVID-19 pandemic on a global level. The UNESCO Awards partnered with NTFCF, which supports five award cycles of 2021–2025, thereby increasing the impact of award programs (UNESCO, 2022). According to the data, UNESCO has also introduced a category of awards for Special Recognition for Sustainable Development to acknowledge the critical role of cultural heritage towards sustainable development.

The global strategy initiated by UNESCO for the World Heritage Convention reflects that 178 countries have signed the strategy in the last 10 years (UNESCO, 2023b). Therefore, the strategy of UNESCO aims at preservation and heritage management through the collective efforts of the countries to preserve the heritage sites for sustainable development.

1.3.2 BRANDING

According to Balmer & Burghausen (2019), a heritage brand is defined by its longevity, positioning, and value proposition. All three of these aspects require targeted marketing and communication. According to Pera (2017) and Yilmaz (2020), the product's quality should mirror that of the location, and the narrative should elicit positive feelings and be interesting to listen to. Therefore, successful delivery of corporate, regional, destination, and individual branding can be said to have been achieved by the country or location whose marketing is successful in appealing to environmental and cultural connections. The heritage of a region is its authentic reflection, which teaches tourists about the region's cultural history and the values that have been passed down through the generations. Marketing campaigns are an efficient technique to communicate the heritage values of a site and to develop an identity for it. In order to successfully attract tourists to heritage places, a well-planned marketing approach is needed. A great number of developed and developing countries have had success in attracting heritage visitors to their countries by successfully utilizing various marketing strategies and methods.

1.4 SOCIAL-CULTURAL ASPECTS OF HERITAGE MANAGEMENT AT A TOURIST DESTINATION

1.4.1 IMPACT OF SOCIO-CULTURAL ASPECTS ON TOURISM

There can be both positive and negative socio-cultural impacts of tourism, which tend to be more subjective and qualitative in nature. As critiqued by Fletcher et al. (2018), all countries that are involved in the tourism business experience sociocultural changes as a consequence of tourism activity. For example, in Spain, the dynamics of sociocultural change have been found to be rapid since the impacts of tourism development in the Costa Brava were demonstrated in the late 1960s and early 1970s (Fletcher et al., 2018). It has been opined that the host population is influenced by the behavior of tourists. People who are influenced by the behavior of others are likely to be influenced by other members of their community by their changed behavior and attitudes. This can be categorized as the indirect socio-cultural impact on tourism development, which is successful in creating new employment opportunities (Figure 1.3).

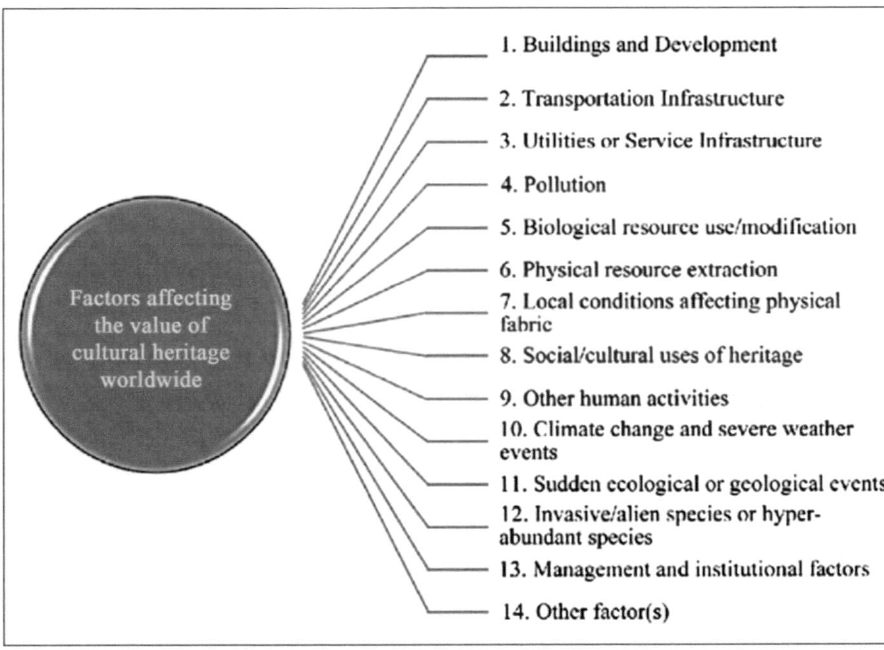

FIGURE 1.3 Factors of the cultural heritage.

Source: Reprinted from Albu & Lesan, (2021) https://creativecommons.org/licenses/by/4.0/.

1.4.2 SOCIETY'S VALUING OF HERITAGE

Heritage is strongly associated with the values, beliefs, and traditions that are told through the stories of place, time, community, and people. Heritage is somewhat linked with social connections and interactions. According to Dans & González (2019), the social dimension of value has become a debate cornerstone on conservation management along with tourism sustainability with regard to cultural heritage. The recently published "Special Eurobarometer on Cultural Heritage" discloses the modern relevance of cultural heritage for European society and economy. In an interview, 80% of Europeans have expressed that cultural heritage is essential to them on a personal level. This shows the higher social value of heritage from the end of the public. Cultural heritage management research helps in developing the dimension of social value as a determining factor in the preservation of cultural heritage. In this case, "social value has also been defined as the collective attachment that represents a set of meanings and values that are most essential to a community." Hence, society's valuing of cultural heritages helps in identifying their desire to preserve or conserve cultural heritage.

1.4.3 INNOVATIVE WAYS OF TRADITIONAL TOURISM DEVELOPMENT

Connecting to a particular place, heritage buildings, and monuments can help in growing a sense of spirituality among the people. Therefore, sticking to the same beliefs and cultures can help in the development of tourism. According to Bozorovich (2019), sustainable tourism along with socioeconomic development is directly associated with the principles of sustainable development. On the other hand, Kayumovich (2020) has opined that in the 21st century, the tourism sector has observed growth in information and communication technologies. However, to retain the old traditional cultural heritage, it can be observed that it is the knowledge sharing and holding on to the old values, customs, beliefs, traditions, and mindset of the people in order to develop the tourism sector massively.

1.4.4 IMPACTS OF TOURISM/VISITOR/RECREATION ON THE ECONOMY

The impact of a large number of tourists and visitors in several countries helps in the effective development of the tourism sector. The evidence shows that GDP growth accelerated due to the travel and tourism sector as "it rose by 21.7% in 2021" yet dropped in 2020 due to the COVID-19 pandemic (Statista, 2023). However, the overall contribution can be estimated to be "5.81 trillion US dollars" indicating a sharp growth (Statista, 2023) (Figure 1.4).

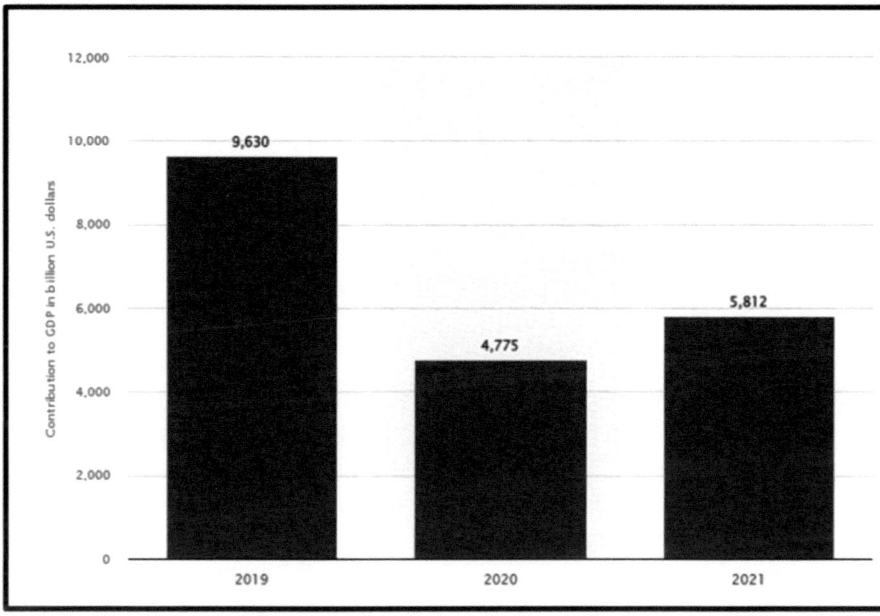

FIGURE 1.4 Global tourism contribution to GDP growth.

Source: Reprinted from Statista (2023) https://www.statista.com/statistics/233223/travel-and-tourism-total-economic-contribution-worldwide/.

Ren et al. (2019) opined that the tourism sector has made a huge contribution to GDP improvement and also plays an important role in terms of creating employment opportunities, reducing poverty, and enhancing income distribution. It has been opined by Comerio & Strozzi (2019) that the tourism industry is now included in three of the "17 universal goals including goal 8 on decent work and economic growth, goal 12 on responsible consumption, and goal 14 on life below water." Hence, its dynamism and contribution to economic growth can be understood through this evaluation.

1.5 FUTURE OF HERITAGE MANAGEMENT AND SUSTAINABLE TOURISM

1.5.1 CONSERVATION AND RESILIENCE OF HERITAGE MANAGEMENT

Conservation of heritage buildings is necessary to provide a sense of continuity and identity in a fast-changing world for future generations. Since heritage buildings resemble the past culture and history of a nation, society, and community, it is fundamental to hold on to the cultural continuity of what constitutes together the architectural heritage of an area. As opined by Bui et al. (2020), tourism resilience helps in identifying key components of resilient tourism such as harnessing leadership, market forces, stakeholder cohesion, and contextual systems such as adaptability, learning, and flexibility. Not only conserving heritage monuments and places is necessary, but it can also be understood that conserving these places can also bring resilience to the global economy. On the other hand, Avrami et al. (2019) have opined that cultural heritages can highly be impacted by pollution and man-made disruptions, which can eventually influence heritages in a negative manner. The author has also discussed the dynamic and constant changes that occur due to feedback and interactions between several parts of the Earth's system. However, climate change is driven by the increase in carbon emissions led to excessive "CO_2 that is more than 410 ppm in 2020," impacting heritage and its values. Hence, this has simultaneously raised the concern for conserving and managing cultural heritages in an effective manner. There is a need to develop heritage tourism that can help improve local socioeconomic conditions while also ensuring long-term viability for future generations. There is a need for utilizing a strategy that is comprehensive, inclusive, multi-sectoral, and multifaceted, integrating every aspect of cultural stakeholders, ranging across the countless caretakers and spectrum of heritage experts to the visitors as well as people of India. There is not one single overarching vision for the heritage of India; therefore, its conservation, preservation, and promotion have not been properly steered in the right direction up until this point. This is simply due to the fact that India's history can be found in a wide variety of forms, shapes, and experiences and that these may be found in many different states. In contrast to the cultural heritage of other Western countries, India has maintained its age-old customs throughout the centuries, making it a unique example of cultural persistence. 'Living Heritage' and this dynamic need to be recognized and managed effectively while developing a management framework for Heritage. This is necessary in order to preserve India's cultural history.

This heritage not only makes up the significant landmarks of India's history but also offers a one-of-a-kind chance for the country to increase its employment rate as well as its income level through the promotion of historical tourism and the growth of local communities. Tourist places are meant to attract tourists and drive the growth and development of the sector. As opined by Liu (2020), heritage conservation is necessary for "a communication act" that is the transmission and inheritance of heritage value. Hence, it is an activity that is focused on raising consumer awareness and improving the public's insights regarding cultural heritage sites as a process of conserving heritage. This is done in order to retain the continuity of the development of the tourism sector which acts as a catalyst for driving growth in the tourism sector. Tourism resilience also lies in the good customer experience of the visitors at the heritage sites. However, "interactive and digital media" were set as the survey scales in Kempiak's investigation of the experience of the visitors in Northern Ireland which played a major role in giving the visitors a good experience. As argued by Fletcher et al. (2018), the sociocultural impacts are highly associated with tourism which can be either positive or negative at times. Nevertheless, one positive impact is underlined by *Kadt*, which is the exchange of cultural ideas, information, and beliefs. Nevertheless, tourism can also be seen as conserving the host's cultural heritage, such as in Petra, York, and Jordan in the UK, and Machu Picchu in Peru. This has a positive impact as it extends over ancient monuments, crafts, sites, rituals, and cultural ceremonies. Hence, all these factors become very important in conserving the heritage of cultural norms and values in different places.

1.5.2 DEVELOPMENT ASPECTS OF SUSTAINABLE TOURISM

"The World Tourism Organization (WTO)" has defined tourism as the way that takes complete account of its current and future economic as well as environmental and social impacts that can help in identifying the requirements of visitors, the sector, and host communities (SDGS, 2023). By 2030, Sustainable Development *SDG targets 8.9* aims to create employment and promote local products and culture. This highlights the cruciality of sustainable development in tourism that aims to integrate and implement tools that can help in monitoring sustainable development impacts for sustainable tourism (SDGS, 2023). As critiqued by Streimikiene et al. (2021), sustainable tourism development is intertwined with paying attention to customers as this helps in attracting more consumers, expand business, and increase competitiveness. In order to mitigate environmental issues, emphasis is put on green tourism to reduce carbon emissions since the tourism sector contributes 8% of the carbon dioxide gas in the environment. Sustainable tourism not only promotes nature preservation or socially responsible business. It is necessary to remember that issues in tourism are an area that receives less social involvement and actions. The overarching goal of World Heritage and the various stakeholders in the tourist industry is to share responsibility for the preservation of our shared cultural and natural heritage that possesses Outstanding Universal Value, as well as for the promotion of sustainable development through the prudent management of tourism.

1.6 CONCLUSION

From the above discussion, it can be analyzed that tourism has contributed the highest GDP to the global economy and it can be stated that it is important to know that tourism is one of the most promising sectors in the global economy. It can be interpreted that managing cultural heritage is one of the sustainable activities in the tourism sector that needs to be maintained for future generations. This is done in order to impart the knowledge, art, culture, values, and beliefs of the people. It can also be assumed that society has huge respect and value for cultural heritages in order to understand the contribution of this particular sector in the global economy.

The stakeholders are making strides toward protecting the built heritage of the country for future generations through the introduction of novel ideas and policies. The chapter highlights the fact that heritage still has a long way to go in terms of environmental and social sustainability, cultural variety, and unified urban expansion.

Cultural preservation is a multifaceted endeavor that calls for input from experts in many disciplines (history, economics, engineering, archaeology, etc.). Increasingly, discussions about heritage management are centering on how to foster sustainable tourism practices. The goal is to provide economic benefits to host communities while minimizing negative impacts on the environment and local wildlife. The realization that our planet's resources are finite and that we must preserve them for the sake of future generations is a major factor in this shift.

To build a rational framework that supports sustainable preservation, it's important to think about the integration of technology more than only the site's physical characteristics, such as its economic and investment potential, environmental impact, social behaviors, and cultural significance. Digital tools and virtual platforms are used for better planning and implementing virtual tours, interactive exhibits, and virtual interactions with stakeholders. The technology is being leveraged to improve the visitor's experience. The growing recognition is fostering a sustainable approach to heritage management and tourism development.

However, the chapter has relied highly on the conservation of cultural heritages to make it value-based for future generations. The critical discussion on sustainable tourism has also been able to underpin the notion and aspects of developed sustainable tourism that add more to the global economy. However, the sector had a setback in the global pandemic, but it also compensated for the loss of visitors. Lastly, it can be evaluated that managing cultural heritage not only helps in managing and conserving heritage buildings or monuments but is also focused on maintaining a socio-cultural balance in the environment.

1.7 THEORETICAL IMPLICATIONS

The introduction paper provides current empirical support in relation to heritage tourism brand distinctiveness, which enables it to make theoretical contributions toward advancing tourism research. A noteworthy theoretical conclusion of this research is the recording of variances and similarities surrounding brand collaboration of heritage tourism with sustainable tourism among tourists. The purpose of this study was to provide useful recommendations for the enhancement of heritage tourism locations by focusing on the behavior of tourists in relation to the authenticity of heritage tourism. It is the responsibility of the local authorities of historical tourism locations to preserve the authenticity of both tangible (such as the architecture) and intangible (such as the local myths and stories) features, and they should also provide activities that elicit a sense of authenticity in visitors in order to encourage return visits. In order to accomplish this, authentic tourism activities need to be designed while keeping in mind the architecture and myths of the surrounding area, as well as taking into account the kinds of activities that visitors are likely to engage in. In addition, tourism activities should assist in educating visitors about the local culture and provide them the opportunity to engage in conversation with locals in a way that is authentic, forthright, and cordial. Heritage tourist locations should work toward unifying the economic and social benefits of their operations in order to preserve and capitalize on their authentic qualities. This research has made a significant contribution to the idea of brand differentiation as well as heritage tourism.

1.8 MANAGERIAL IMPLICATIONS

The tourism stakeholders need to create and implement sustainable development strategies that strike a balance between the protection of cultural landmarks and the economic and societal gains that may be made from increased heritage tourism. It is necessary to network with community members, heritage site protectors, government agencies, non-governmental organizations, and others involved in the project. The interests and concerns of all parties are more likely to be addressed in inclusive and participatory decision-making processes.

Most cultural heritage sites rely on governmental funding, and this is likely to continue; however, it is unrealistic to anticipate receiving unlimited funding. When funding for heritage sites and other cultural institutions is so dependent on government grants, the future of these institutions is always up in the air. It would be unreasonable to anticipate that most locations would suddenly become financially self-sufficient or that managers would have an entrepreneurial spirit. The improvement in financial management will lead to increased sustainability, and financial guarantees that it will be preserved and cared for future generations. The findings may provide important clues for managers looking to boost heritage sites' economic and social viability. The concept that providing technological and organizational innovations improves economic and social sustainability is strengthened by the fact that innovativeness has an effect on tourism's long-term viability. Furthermore, it has been demonstrated that a focus on the market is a prerequisite for ensuring long-term economic and social viability.

KEYWORDS

- **cultural heritage management (CHS)**
- **gross domestic production (GDP)**
- **heritage**
- **heritage management**
- **sustainable development goals (SDGS)**
- **sustainable practices**
- **The International Council on Monuments and Sites (ICOMOS)**
- **The World Tourism Organization (WTO)**
- **tourism**
- **United Nations Educational**
- **Scientific and Cultural Organization (UNESCO)**

REFERENCES

Albu, S., & Lesan, A. (2021). Expropriation as a way for protecting the architectural heritage. *Current Urban Studies*. Advance online publication. https://doi.org/10.4236/cus.2021.94029.

Avrami, E., Macdonald, S., Mason, R., & Myers, D. (Eds.). (2019). *Values in Heritage Management: Emerging Approaches and Research Directions*. Oxford University Press.

Balmer, J. M., & Burghausen, M. (2019). Marketing, the past and corporate heritage. *Marketing Theory, 19*(2), 217–227. https://doi.org/10.1177/1470593118822766.

Blasco López, M. F., Recuero Virto, N., Aldas Manzano, J., & Garcia-Madariaga, J. (2018). Tourism sustainability in archaeological sites. *Journal of Cultural Heritage Management and Sustainable Development, 8*(3), 276–292. https://doi.org/10.1108/JCHMSD-09-2016-0051.

Boniotti, C. (2023). The public–private–people partnership (P4) for cultural heritage management purposes. *Journal of Cultural Heritage Management and Sustainable Development, 13*(1), 1–14. https://doi.org/10.1108/JCHMSD-05-2022-0031.

Bozorovich, U. C. (2019). Innovative ways of traditional tourism development. *Asian Journal of Technology & Management Research (AJTMR), 2249*(0892), 84–87.

Bui, H. T., Jones, T. E., Weaver, D. B., & Le, A. (2020). The adaptive resilience of living cultural heritage in a tourism destination. *Journal of Sustainable Tourism, 28*(7), 1022–1040. https://doi.org/10.1080/09669582.2019.1695554.

Capetownmagazine. (2023). Top adventures and extreme sports in Cape Town. Retrieved from: https://www.capetownmagazine.com/extreme-sports (accessed on 18 June 2024).

Comerio, N., & Strozzi, F. (2019). Tourism and its economic impact: A literature review using bibliometric tools. *Tourism Economics, 25*(1), 109–131. https://doi.org/10.1177/1354816618768198.

Dans, E. P., & González, P. A. (2019). Sustainable tourism and social value at World Heritage Sites: Towards a conservation plan for Altamira, Spain. *Annals of Tourism Research, 74*, 68–80. https://doi.org/10.1016/j.annals.2018.10.006.

Economictimes. (2023). Tomatina Festival Spain. Retrieved from: https://economictimes.indiatimes.com/topic/la-tomatina (accessed on 18 June 2024).

Elshater, A. (2019). The predicament of post-displacement amidst historical sites: A design-based correlation between people and place. *Heritage & Society, 12*(2–3), 85–115. https://doi.org/10.1080/2159032X.2019.1601931.

Eppich, R., & Grinda, J. L. G. (2019). Sustainable financial management of tangible cultural heritage sites. *Journal of Cultural Heritage Management and Sustainable Development, 9*(3), 282–299. https://doi.org/10.1108/JCHMSD-11-2018-0081.

Fletcher, J., Fyall, A., Gilbert, D., & Wanhill, S. (2018). *Tourism: Principles and Practice* (6th ed.). Pearson.

Foster, G. (2020). Circular economy strategies for adaptive reuse of cultural heritage buildings to reduce environmental impacts. *Resources, Conservation and Recycling, 152*, Article 104507. https://doi.org/10.1016/j.resconrec.2019.104507.

Guzman, P. (2020). Assessing the sustainable development of the historic urban landscape through local indicators: Lessons from a Mexican World Heritage City. *Journal of Cultural Heritage, 46*, 320–327. https://doi.org/10.1016/j.culher.2020.03.002.

ICOMOS. (1996). Declaration of San Antonio. Retrieved from: https://www.icomos.org/en/resources/charters-and-texts/179-articles-en-francais/ressources/charters-and-standards/188-the-declaration-of-san-antonio (accessed on 18 June 2024).

ICOMOS. (1998). The Stockholm Declaration: Declaration of ICOMOS marking the 50th anniversary of the Universal Declaration of Human Rights. Retrieved from: https://www.getty.edu/conservation/publications_resources/research_resources/ charters/charter67.html.

Kayumovich, K. O. (2020). Prospects of digital tourism development. *Economics, 1(44),* 23–24.

Liu, Y. (2020). Evaluating visitor experience of digital interpretation and presentation technologies at cultural heritage sites: A case study of the old town, Zuoying. *Built Heritage, 4*(1), 14. https://doi.org/10.1186/s43238-020-00006-9.

Mayer, A. L. (2008). Strengths and weaknesses of common sustainability indices for multidimensional systems. *Environment International, 34*(2), 277–291. https://doi.org/10.1016/j.envint.2007.06.004.

Miller-Rushing, A. J., Athearn, N., Blackford, T., Brigham, C., Cohen, L., Cole-Will, R., Edgar, T., Ellwood, E. R., Fisichelli, N., Pritz, C. F., & Gallinat, A. S. (2021). COVID-19 pandemic impacts on conservation research, management, and public engagement in US national parks. *Biological Conservation, 257*, Article 109038. https://doi.org/10.1016/j.biocon.2021.109038.

Nbusinessinfo. (2023). *Niche Tourism Opportunities*. Retrieved from: https://www.nibusinessinfo.co.uk/content/activities-and-nature-based-tourism#:~:text=hard%20adventure%20activities%20such%20as,tours%2C%20canoe%20hire%20or%20camping (accessed on 18 June 2024).

Nilsson, M., Chisholm, E., Griggs, D., Howden-Chapman, P., McCollum, D., Messerli, P., & Stafford-Smith, M. (2018). Mapping interactions between the sustainable development goals: Lessons learned and ways forward. *Sustainability Science, 13*, 1489–1503. https://doi.org/10.1007/s11625-018-0624-5.

Orr, S. A., Richards, J., & Fatorić, S. (2021). Climate change and cultural heritage: A systematic literature review (2016–2020). *The Historic Environment: Policy & Practice, 12*(3-4), 434–477. https://doi.org/10.1080/17567505.2021.1958675.

Pera, R. (2017). Empowering the new traveler: Storytelling as a co-creative behavior in tourism. *Current Issues in Tourism, 20*(4), 331–338. https://doi.org/10.1080/13683500.2015.1064845.

Ren, T., Can, M., Paramati, S. R., Fang, J., & Wu, W. (2019). The impact of tourism quality on economic development and environment: Evidence from Mediterranean countries. *Sustainability, 11*(8), Article 2296. https://doi.org/10.3390/su11082296.

Richards, J. (2018). *What is Heritage Management?* Retrieved from: https://blogs.exeter.ac.uk/humanities/2018/03/29/what-is-heritage-management/#:~:text=The%20practice%20of%20heritage%20management,the%20marketing%20of%20heritage%20sites (accessed on 18 June 2024).

Sabou, G. C. (2012). Sustainable Heritage Management – A Matter of National Policies Congruence. *International Journal of Economic Practices and Theories, 2*(3), 137–142.

SDGS. (2023). *Sustainable tourism*. Retrieved from: https://sdgs.un.org/topics/sustainable-tourism#:~:text=The%20World%20Tourism%20Organization%20defines,the%20environment%20and%20host%20communities%22 (accessed on 18 June 2024).

Statista. (2023). *Total Contribution of Travel and Tourism to Gross Domestic Product (GDP) Worldwide from 2019 to 2021*. Retrieved from: https://www.statista.com/statistics/233223/travel-and-tourism--total-economic-contribution-worldwide/ (accessed on 18 June 2024).

Streimikiene, D., Svagzdiene, B., Jasinskas, E., & Simanavicius, A. (2021). Sustainable tourism development and competitiveness: The systematic literature review. *Sustainable Development, 29*(1), 259–271. https://doi.org/10.1002/sd.2170.

Sun, H. (2010). World Heritage Classification and Related Issues—A Case Study of the "Convention Concerning the Protection of the World Cultural and Natural Heritage. *Procedia Social and Behavioral Sciences, 2*(5), 6954–6961. https://doi.org/10.1016/j.sbspro.2010.05.095.

UNESCO. (1972). Convention concerning the Protection of the World Cultural and Natural Heritage 1972. Paris: UNESCO World Heritage Centre.

UNESCO. (2016). Operational Guidelines for the Implementation of the World Heritage Convention. Paris: UNESCO World Heritage Centre.

UNESCO. (2021a). Impacts of COVID-19 on World Heritage to continue for years, according to report. Retrieved from: https://whc.unesco.org/en/news/2298 (accessed on 18 June 2024).

UNESCO. (2021b). United nations educational, scientific and cultural organization convention concerning the protection of the world cultural and natural heritage. Retrieved from: https://whc.unesco.org/archive/2021/whc21-44com-inf5A.2-en.pdf (accessed on 18 June 2024).

UNESCO. (2021c). *Socio-Economic Impacts of World Heritage Listing.* Retrieved from: https://whc.unesco.org/en/socio-economic-impacts/ (accessed on 18 June 2024).

UNESCO. (2022). *2022 UNESCO Asia-Pacific Awards for Cultural Heritage Conservation: Announcement of Winners*. Retrieved from: https://bangkok.UNESCO.org/content/2022-unesco-asia-pacific-awards-cultural-heritage-conservation-announcement-winners#:~:text=BANGKOK%2C%2026%20Nov%202022%20%E2%80%93%20The,Pacific%20Awards%20for%20Cultural%20Heritage.

UNESCO. (2023a). *World Heritage and Sustainable Development.* Retrieved from: https://whc.unesco.org/en/sustainabledevelopment/ (accessed on 18 June 2024).

UNESCO. (2023b). *Global Strategy*. Retrieved from: https://whc.unesco.org/en/globalstrategy/ (accessed on 18 June 2024).

Uzzell, D. L. (1996). Creating place identity through heritage interpretation. *International Journal of Heritage Studies, 1*(4), 219–228. https://doi.org/10.1080/13527259608722146.

VisitScotland. (2023). *Top Tourist Attractions & Places to Visit in Scotland.* Retrieved from: https://www.visitscotland.com/things-to-do/attractions/most-visited (accessed on 18 June 2024).

Vu, H. Q., Li, G., & Law, R. (2020). Cross-country analysis of tourist activities based on venue-referenced social media data. *Journal of Travel Research, 59*(1), 90–106. https://doi.org/10.1177/0047287519871991.

WTTC. (2023). *Economic Impact Reports.* Retrieved from: https://wttc.org/research/economic-impact (accessed on 18 June 2024).

Yilmaz, E. S. (2020). The effects on consumer behavior of hotel related comments on the Tripadvisor website: An Istanbul case. *Advances in Hospitality and Tourism Research (AHTR), 8*(1), 1–29.

CHAPTER 2

INTERPRETATION AS A MANAGEMENT TOOL FOR BIRDWATCHING TOURISM SITES: THE CASE OF THE ARENA BLANCA RESERVE, ALTO MAYO, PERÚ

PABLO CÉSAR ROMO-ROMÁN,[1]
ROSSE MARIE ESPARZA-HUAMANCHUMO,[2,3]
MARÍA DE LA CRUZ DEL RIO-RAMA,[4] and
JOSÉ ÁLVAREZ-GARCÍA[5]

[1]*Faculty of Business Sciences of the Universidad San Ignacio de Loyola, Lima, Perú*

[2]*Research Group in Gastronomy and Culinary Art, Faculty of Hotel Administration, Tourism and Gastronomy, Universidad San Ignacio de Loyola, Lima, Peru*

[3]*Doctoral Program in Sustainable Territorial Development (R015), International Doctoral School, University of Extremadura, Badajoz, Spain*

[4]*Business Management and Marketing Department, Faculty of Business Sciences and Tourism, University of Vigo, Ourense, Spain*

[5]*Departamento de Economía Financiera y Contabilidad, Instituto Universitario de Investigación para el Desarrollo Territorial Sostenible (INTERRA), Universidad de Extremadura, Cáceres, Spain*

Dynamics of Heritage Management and Tourism Development: Trends, Practices, and Approaches.
Surabhi Srivastava & Ruchika Kulshrestha (Eds.)

ABSTRACT

Bird watching is a tourist activity that has been growing steadily in recent years. Its specialization requires the implementation of various activities to achieve a high value-added offer. Peru is a country with enormous potential for birdwatching tourism, specifically the San Martín region is one of the privileged places. It is part of the northern bird circuit of Peru with a great diversity of birds from the most common to the endemic, where this type of specialized tourism can be developed. The present investigation is carried out in the Arena Blanca Ecological Reserve, a site specialized in providing bird-watching services located in Alto Mayo, San Martín, Peru. The objective is to analyze the implementation of continuous heritage interpretation processes that facilitate bird-watching, as well as the management of the tourist site. This research has a mixed methodological basis. The qualitative approach will allow detection of the implementation process of the interpretation as a management tool for the tourist site for observing wild birds. Likewise, the approach will allow collecting data to evaluate the effectiveness and satisfaction of the operators, tourists, guides, and/or visitors in the visiting area. The analysis is intended to demonstrate that the processes of interpretation of the heritage allow for facilitating the observation of wild species and how it contributes to the management of the tourist site.

2.1 INTRODUCTION

Heritage interpretation began in the late 19th century with the creation of the world's first national park, Yellowstone, in 1872, to guide visitors to protected natural areas in the wilderness. Therefore, following a growing trend for the conservation and creation of more protected areas such as Sequoia, Yosemite, and Rainier national parks between 1890 and 1900, and the creation of the National Park Service in 1916, "heritage interpretation" appeared, initially conceived as a way of guiding in nature (Brunelli, 2019).

By 1920, the activities of guiding tourists visiting national parks had become extended to the point that an Education Division was created in 1923, which promoted the creation of nature guides and interpreters. It should be noted that throughout this process, personalities such as John Muir, Enos Mills, and Freeman Tilden, among others, emerged (Brunelli, 2019). Currently, heritage interpretation has spread and is seen as an educational process, mostly informal, that contributes to raising awareness among visitors, local people, and decision-makers, among others, regarding the importance of the place, the resources it has, and the conservation of nature in general.

In addition, interpretation is used as a process that contributes to the management of natural tourism sites, as it can contribute to reducing the negative impacts of tourist visits because it contributes to the management of groups and the flow of tourists. It also increases the environmental awareness of visitors and therefore manages their behavior and captures their interest in nature conservation (Mearns & Botha, 2018).

In countries like Peru, with a high level of biodiversity and a variety of ecosystems, different areas have a high potential for developing various tourism activities based on biological and natural diversity. One of these activities is bird watching. In this respect, Peru is the country with the second-largest number of birds in the world (Birdlife, 2023), with up to 1884 bird species recorded (Plenge, 2022). This activity has a high potential to generate economic, social, and environmental benefits.

In this line, some territories of the country are developing small clusters related to bird watching, where the basis for the development of these economic clusters is the observation sites that, in addition to being dedicated to conservation work, provide services for observing birds. These services facilitate bird watching, increase the success of sightings, and have processes that sensitize visitors to the importance of conservation of the species and the ecosystems where they live.

In this sense, these areas specialized in providing birdwatching services need to implement adequate planning processes to improve their services and manage the impacts of tourism properly, an aspect highly valued by the international birdwatching tourism market. Within the necessary planning processes, adequate heritage interpretation is crucial to ensure the quality of the services provided by the Reserve and to contribute to ensuring its proper environmental management and contribution to biodiversity conservation.

This chapter analyzes the interpretive proposal implemented by the Arena Blanca Ecological Reserve, the perception of the value chain made up of tour operators, official tour guides, and biologists who are in charge of conducting research in the area, among other key actors who contribute to this place. It also aims to measure the satisfaction of the Arena Blanca Ecological Reserve and, based on this, contribute to improving its interpretive proposal.

2.2 LITERATURE REVIEW

2.2.1 HERITAGE INTERPRETATION

According to Tilden (1977), heritage interpretation is "an educational activity that seeks to reveal meanings and interrelationships through the use of original objects, by direct contact with the resource, or by illustrative means, not merely providing factual information." On the other hand, Morales and Ham (2008) point out that "Effective interpretation is a creative process of strategic communication, which produces intellectual and emotional connections between the visitor and the resource being interpreted, making them generate their meanings about that resource so that they can appreciate and enjoy it."

In addition, the Asociación para la interpretación del Patrimonio (n.d.) states that heritage interpretation is the art of revealing the natural and/or cultural heritage to the public visiting a given site. It also indicates that effective interpretation can function as a management tool, as it reduces the need for regulatory compliance, as well as reducing maintenance costs, assists in the management of the flow of people and visitor areas, and can be a means of gaining public support for the interpretive agency providing interpretation. For Hongyu et al. (2022), governance plays a crucial role in achieving sustainability goals in destinations that conserve their cultural and natural heritage.

In the arts, for example, the actors or artists are the transmitters and the public are the receivers; how the message is transmitted, the elements they use, the personality, and others, are part of a set of symbolic elements that the transmitter(s) use to provoke a reaction in the receiver and at the same time generate analysis and share ideas. In the case of heritage interpretation, behind each interpretive medium or element, there is an institution that it serves and that is collectively recognized. Then, the concept acquires a particular relevance because it possesses the intention to generate changes in attitudes or states among those who receive the message (Dumbraveanu, Craciun, & Tudoricu, 2016).

Tilden (1977) also establishes six principles of heritage interpretation, which are: (1) all interpretation must relate to the personality or previous experience of the visitor, (2) pure information is not interpretation, interpretation is the revelation of a situation based on information, (3) interpretation is an art, combining many arts, (4) the main objective of interpretation is provocation, (5) interpretation must present a whole rather than apart, and (6) interpretation aimed at children must implement a differentiated program.

2.2.2 TOURISM BASED ON BIRDWATCHING

Wildlife watching is a very popular activity (Orams, 2002) that is related to the scientific study of nature (Dobson, 2008) and is part of nature-based tourism. Among the main characteristics, the direct contribution to nature conservation stands out, due to environmental education processes for visitors and local populations (Apps, Dimmock, & Huveneers, 2016), as well as the generation of financial funds to implement conservation actions.

In this regard, tourism based on wildlife observation is associated with an animal species or a group of species, especially buffaloes, elephants, leopards, lions, rhinoceroses, gorillas, lemurs, birds, whales, and dolphins (UNWTO, 2015). Safaris stands out as wildlife-related expeditions that are associated with hunting, observation, and photography of wild species and are mainly conducted in protected areas. Safaris include a range of activities and services including lodges and camps, four-wheel drive vehicles suitable for observing and photographing wildlife, and recently, other modalities of mobilization or related activities such as trekking, canoeing, or camel safaris are emerging (UNWTO, 2015).

Birdwatching is a practice that consists of observing and identifying birds in their natural habitats by eye and ear (Sekercioglu, 2002). It is an activity related to wildlife watching and nature-based tourism, and it shows global growth (Connell, 2009). Birdwatching can function as a complementary tourism activity or be the main activity or attraction of a journey.

According to the United States Fish & Wildlife Service (2016), birdwatching is the leading wildlife-watching activity in the United States, with approximately 45 million birdwatchers. It is one of the most popular outdoor recreational activities in Western countries (Ma et al., 2013). In addition, it has started to become more popular in other latitudes, especially in remote areas where rare bird species can be found (Connell, 2009), as is the case in South America in countries like Peru, Colombia, and Ecuador.

According to Schwoerer & Dawson (2022), birdwatchers have increased in recent years and could be said to be the largest group of ecotourists in the world. In this regard, tourism based on birdwatching, like other activities, generates interaction between tourists and the community and requires a healthy state of conservation of the ecosystems that comprise the destination and on which the quantity, variety, and particularity of bird species depend. This is the main reason for birdwatchers to travel.

In general, it is noted that birdwatchers who travel have a higher-than-average educational level, adequate economic income, and are highly committed to birdwatching (Schwoerer & Dawson, 2022); however, in recent years they have become an increasingly heterogeneous community. In addition, concerning the reasons for deciding to visit a birding destination, the most important are the characteristics related to the presence or particularity of birds, such as endemism, endangered species, migration, courtship behavior, and diversity, among others; additionally, and complementarily, the accessibility of the destination, guide, lodging, transportation, and affordable prices were identified (Stemmer, Aas, Veisten, & Lindberg, 2022).

According to Chen & Chen (2015), the most important factors for visiting a birdwatching destination are the presence of special or never-seen birds, and the availability of professional and capable guides, among other attributes. However, reasonable cost proximity and accessibility could be more important than bird-related factors, affecting their decision-making.

Finally, it is necessary to indicate that there is enough evidence that determine the potential of the tourism activity based on birdwatching, especially for those countries that show a high biodiversity of birds and should design and implement products with high service standards to achieve insertion in the international birdwatching market. Peru is one of the countries with the largest number of birds in the world, with 1891 recorded species, 117 of which are endemic (Plenge, 2023).

2.3 METHODOLOGY

2.3.1 STUDY SITE

The Arena Blanca Ecological Reserve is a private property belonging to Norbil Becerra Garcia, located in the village of Aguas Verdes in the buffer zone of the Protected Natural Area, Alto Mayo Protected Forest, district of Pardo Miguel, province of Rioja, department of San Martin, on the edge of the Serranoyacu River.

The Reserve began operations in August 2013 in a one-hectare area, and currently, the area covered by the Reserve is 16 hectares of conserved forest. Additionally, it is located 1,146 meters above sea level, in a very humid premontane tropical forest ecosystem settled on white sand that has reduced organic matter and low nutrients, which retains little water and suffers greater drought, and therefore generates vegetation that atrophies and has unique plants and animals adapted to that ecosystem.

From Tarapoto, a city with air connections from Lima, you can get to the Arena Blanca Reserve by following the Belaunde Terry highway towards Pedro Ruiz. From Tarapoto, you can get there in 4 hours, and from Moyobamba in approximately 2 hours and a half (Figure 2.1).

Arena Blanca Reserve is strategically located in an Important Bird Area – IBA, Abra Patricia – Alto Mayo (Birdlife, 2023), as well as being in a transition zone between the Andes and the Amazon, with high biodiversity and endemism, particularly for birds.

FIGURE 2.1 The Arena Blanca Ecological Reserve.

Young et al. (2007) have identified the areas with the greatest bird endemism on the eastern slopes of the Andes and have indicated that the departments of Cusco and Huánuco in Peru have the greatest number of endemic species per km^2 and the areas between the Colan and Alto Mayo Mountain ranges, in the departments of Amazonas and San Martín, have endemic species with the smallest distributions.

In addition, according to Puhakka, Salo & Saaksjarvi (2011), there are 5 areas in the country with potential for tourism development based on bird watching: southeastern Peru (Cusco and Madre de Dios), central Peru, northern Peru, and areas around Piura and Iquitos. Among these, southeastern Peru stands out for the opportunity to see many species, in the case of central Peru is also recognized by the large number of species and presence of endemics, northern Peru stands out as an area with a high number of endemic bird species with a high degree of restriction or reduced distribution, which makes it very special and attractive to the international market.

In this line, Promperu (2019) has defined 3 bird routes nationally: the Southern Route, the Central Route, and the Northern Route. In the case of the Northern Route, there are two main sections, including the Lambayeque-Tarapoto Road and Pedro Ruíz–Cajamarca Road, the first section of which Reserva Arena Blanca belongs.

It is important to note that this route is one of the most biodiverse in the world because of the variety of ecosystems and geographic complexity, which allows for the presence of a wide variety of bird species, many of which are endemic (Promperu, 2015).

2.3.2 *RESEARCH INSTRUMENT*

The study has a quantitative approach, based on participant observation, and began with a field visit to identify and describe the process of interpretive services offered by the Arena Blanca Ecological Reserve; then, the opinions of the main actors in the value chain of the Arena Blanca Ecological Reserve on the heritage interpretation services they offer were recorded.

In this sense, through a survey method, a specific questionnaire was applied on satisfaction with the heritage interpretation service and related issues in January and February 2023.

The instrument for this research was a questionnaire made up of 21 items, distributed as follows (interpretative resources – 6, research and follow-up – 4, available information and orientation – 5, facilities – 4, and clarity of the interpretive message and contribution to the conservation – 2), a Likert-type scale was used to collect the opinion of the population on heritage interpretation services.

The choice of the questionnaire items was based on the field visit, choosing items related to the context of the reserve. The instrument was applied to 21 members of the value chain related to the Arena Blanca Ecological Reserve who provide tourist services and products such as travel agents, official tourist guides, biologists, and researchers, among others. The data was processed using the SPSS Statistical software, version 22.0.

2.4 DESCRIPTION OF THE HERITAGE INTERPRETATION OFFER IN THE ARENA BLANCA ECOLOGICAL RESERVE

The main service provided by Arena Blanca Ecological Reserve is bird watching. The Ebird platform (n.d.) has a record of 480 bird species. Among its services are the bird observatories that have been adapted to ensure the observation of some of the species found in this space. According to Norbil Becerra, manager of the site, the following activities have been required: (1) identification and registration of birds, (2) analysis of species behavior, (3) selection of species for management, (4) actions for species management, (5) adaptation of facilities, and (6) generation of information.

The following should be considered: (1) identification and registration of birds – this activity is used to map and know the type of bird found in the area, this is one of the main activities for birdwatchers, (2) behavior analysis of species – aims to establish the behavior of the bird where it feeds, how it develops, how it breeds its offspring, etc., (3) selection of species for management – this activity helps to select the species that have been most used in the area, (4) actions for species management – contribute to the care and protection of birds, (5) adaptation of facilities – this activity aims to collaborate with areas such as feeders, rest areas, nesting areas, among others, and (6) generation of information – counting on the above, contribute to having real and updated information that allows it to be used for the benefit of the conservation and sustainability of the species in the area.

Currently, the reserve has three observatories: a Quail and Tinamou observatory, a Hummingbird observatory, and a Tanager observatory. These observatories have recorded up to 50 species of birds and ensure the successful observation of an interesting number of species at any time. This has allowed the observation of species, in this case, birds in the Arena Blanca Reserve, to have grown in recent years for research and/or recreation purposes.

In the first one, the Quails and tinamous observatory, it is possible to observe the Rufous-breasted Wood-Quail, Little and Cinereous Tinamou, Orange-billed Sparrow, and the White-tipped Dove among other species. These species are characterized by spending a lot of time on land and are very difficult to see. Based on this record, work began to take care of the forest, identifying the route of some of these species and providing seed corn.

After several months of constant monitoring and tracking of some species, a site was identified, which was established as a feeder for the species mentioned above, and a hide was built to observe these species because they are very sensitive and would move away at any noise or movement. Currently, the observatory ensures the successful sighting of at least 4 species per day at specific times, especially very early in the morning (from 7 to 9 am).

The second observatory is the hummingbird observatory, which includes a large flower garden attractive to hummingbirds, nectar feeders, and an observatory elevated above the level of the flowers to observe the hummingbirds. Up to 33 species of hummingbirds have been recorded in this observatory. The hummingbird species registered are White-necked Jacobin, Rufous-breasted Hermit, Reddish Hermit, Green Hermit, Black-throated Hermit, Blue-fronted Lancebill, Brown Violetear, Sparkling Violetear, Wire-crested Thorntail, Rufous-crested Coquette, Violet-fronted Brilliant, Long-billed Starthroat, White-bellied Woodstar, Little Woodstar, Amethyst Woodstar, Blue-tailed Emerald, Gray-breasted Sabrewing, Fork-tailed Woodnymph, Many-spotted Hummingbird, Golden-tailed Sapphire, Sapphire-spangled Emerald, Rufous-throated Sapphire, Buff-tailed Sicklebill, Gray-chinned Hermit, Great-billed Hermit, Green-fronted Lancebill, Lesser Violetear, Black-throated Mango, Speckled Hummingbird, Chestnut-breasted Coronet, Black-throated Brilliant, Violet-headed Hummingbird, and Napo Sabrewing (Arena Blanca Ecological Reserve, s.f). It is important to note that this is the first observatory developed in the Reserve.

About hummingbirds: they are a family that only inhabits the Americas, mainly in the tropical zone, and are the smallest birds in the world, weighing between 2 and 20 grams, with long, narrow beaks and small wings. They have a backward and upside-down flight, and they are the only vertebrates capable of floating, with high maneuverability, speed, and endurance. In addition, their feathers stand out for their brightness and metallic coloration, which changes according to the reflection of light. Usually, the males are the most colorful with more striking shapes, due in part to sexual dimorphism (Smithsonian's National Zoo & Conservation Biology Institute, n.d.).

The third observatory is for tanagers, which consists of two zones. One is for the care and reforestation of tree species that produce attractive fruits for birds such as miconias, and another area where a feeder has been installed to provide fruits that are liked by this family of birds. In the same area, a hiding place has been installed to observe them without generating disturbances and thus observe and photograph them with greater time and success. Up to 61 species of tanagers have been recorded in the reserve. The species registered in the feeder are Black-bellied Tanager, White-lined Tanager, Burnished-buff Tanager, Magpie Tanager, Blue-gray Tanager, Palm Tanager, Golden-bellied Euphonia, Orange-bellied Euphonia, Yellow-tufted Woodpecker, Sickle-winged Guan, Buff-throated Saltator, Epaulet Oriole, Orange-billed Sparrow, and Yellow-rumped Cacique.

Until the field visit and interview with the manager of the Reserve, Mr. Norbil Becerra Garcia in January 2023, it was identified that in addition to the implementation of the observatories mentioned above, there was access to the bird register of the Reserve, a brochure of the main 30 birds of the Reserve, a guide to the birds of Peru and in each observatory Norbil provides updated information on the constant monitoring of the species of each observatory within the Reserve. In addition, in most cases, visitors are accompanied by a guide or tour conductor specialist in the field, who provides specialized information on the species of each observatory and other species available in the Reserve and surrounding areas, outside the observatories. It should be noted that there is no fixed physical information available in the observatories, although there is a record of bird species available in Ebird, this is general and not specific to each observatory, nor is it developed solely by the Reserve managers. In addition, although there is a management and monitoring of the main species of each observatory, it is not very formal, there is no notebook or record book or checklist, there is not enough systematization, and there is reduced creation of content with the data available.

2.5 RESULTS AND DISCUSSION

From the 5 dimensions evaluated, clarity of the interpretive message and contribution to conservation, interpretive resources are the most highly valued, followed by research and monitoring, and facilities. The least valued is information available and guided.

2.5.1 *INTERPRETIVE RESOURCES DIMENSION*

In this dimension, the highest score refers to the identification that birdwatching is the main activity occurring in the Ecological Reserve (M=4.75, SD=0.444). Birdwatching success is also high in general (M=4.25, SD=0.716) and particularly in the hummingbird observatory (M=4.55, SD=0.605), which is the oldest observatory that has been implemented and where the Wire-crested Thorntail, which is the image of the reserve's logo, stands out (Figure 2.2).

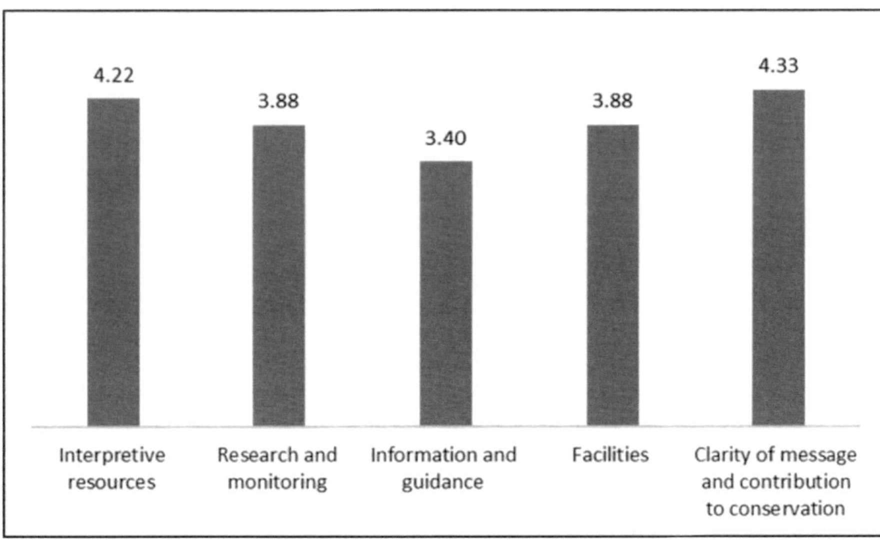

FIGURE 2.2 Mean of the dimensions evaluated.

The lowest scores refer to the success of observation at the Quail and Tinamou observatory (M=3.70, SD=0.716), and the Tanager observatory (M=3.70, SD=0.733). In this regard, it should be noted that in the case of the Tanager observatory, it is the last one implemented and in the case of the Tinamou and Quail observatory, the observation of species is restricted to specific times and sometimes it is not possible to see the respective bird species due to their natural characteristics (Table 2.1).

TABLE 2.1 Descriptive Analysis of Interpretative Resources Dimension

Interpretative Resources Dimension	Mean	SD
1. Is birdwatching the main activity in the reserve?	4.75	0.444
2. Do you agree that the main interpretive resources in the reserve are birds?	4.35	0.587
3. Do you consider the level of birding success in general in the reserve?	4.25	0.716
4. Do you consider satisfactory the level of birding success at the Quail and Tinamou Observatory?	3.70	0.979
5. Do you consider satisfactory the level of birding success at the hummingbird observatory?	4.55	0.605
6. Do you consider satisfactory the level of birding success in the Tanager observatory?	3.70	0.733

2.5.2 *RESEARCH AND MONITORING DIMENSION*

Concerning the research and monitoring dimension, the most highly valued score is the level of recording and monitoring of the hummingbird observatory (M=4.35, SD=0.671); however, the perception of research and monitoring in general in the reserve is lower (M=3.75, SD=0.786), as well as in the quail and tinamou observatory (M=3.70, SD=0.657), as well as in the tanager observatory (M=3.70, SD=0.657) (Table 2.2).

TABLE 2.2 Descriptive Analysis of Research and Monitoring Dimension

Research and Monitoring Dimension	Mean	SD
7. Do you consider satisfactory the level of recording and monitoring of birds in general in the reserve?	3.75	0.786
8. Do you consider satisfactory the level of bird recording and monitoring at the Quail and Tinamou observatory in the reserve?	3.70	0.657
9. Do you consider satisfactory the level of bird recording and monitoring at the hummingbird observatory in the reserve?	4.35	0.671
10. Do you consider satisfactory the level of bird recording and monitoring at the Tanager observatory in the reserve?	3.70	0.657

2.5.3 *INFORMATION AND GUIDANCE DIMENSION*

This dimension is the one with the lowest score. The highest score in this dimension refers to guiding (M=3.60, SD=0.821), which in most cases is done by guides or birding experts from outside the reserve. The lowest scores are for the information available at the Tanager observatory (M=3.20,

SD=0.696), the reserve in general (M=3.30, SD=0.733), and the quail and tinamou observatory (M=3.35, SD=0.745) (Table 2.3).

TABLE 2.3 Descriptive Analysis of Information and Guidance Dimension

Information and Guidance Dimension		Mean	SD
11.	Do you consider that the information provided by the Reserve about birds and related topics is sufficient?	3.30	0.733
12.	Do you consider that the information provided by the Quail and Tinamou observatory is sufficient?	3.35	0.745
13.	Do you consider the information provided by the hummingbird observatory to be sufficient and appropriate?	3.55	0.686
14.	Do you consider that the information provided by the tanager observatory is sufficient?	3.20	0.696
15.	Do you agree that guiding is satisfactory in the reserve? You may consider the work of the external guide and the Reserve staff and/or both.	3.60	0.821

2.5.4 FACILITIES DIMENSION

In general, the results are satisfactory, especially the contribution of the hummingbird observatory facilities (M=4.05, SD=0.887) and the tanager observatory (M=3.90, SD=0.718). The lowest score refers to the contribution of the general facilities in the reserve, which in addition to the observatories could include informative and interpretive signage that coincides with the previous dimension.

It is important to point out that the main services for birdwatching are centered in these facilities and it is the differentiation that exists between these sites and natural areas without birdwatching facilities (Table 2.4).

TABLE 2.4 Descriptive Analysis of Facilities Dimension

Facilities Dimension	Mean	SD
16. Do you consider that the facilities provided by the Reserve contribute to the interpretation services offered by the Reserve?	3.75	0.639
17. Do you consider that the Quail and Partridge Observatory facilities contribute to heritage interpretation?	3.80	0.768
18. Do you consider that the hummingbird observatory facilities contribute to heritage interpretation?	4.05	0.887
19. Do you consider that the tanager observatory facilities contribute to heritage interpretation?	3.90	0.718

2.5.5 CLARITY OF MESSAGE AND CONTRIBUTION TO CONSERVATION DIMENSION

This is the most highly valued dimension. In this sense, the awareness that the reserve generates in its visitors about the conservation of biodiversity (M=4.65, SD=0.587) is high, and the existence of a clear, relevant, and articulated interpretive message between the existing activities and bird observatories (M=4.00, SD=1.076) is also identified.

The results of this dimension are in line with the alliance and the conservation agreement it has with the administration of the Alto Mayo Protected Area, its growth, and contribution to the conservation of this protected area and the conservation of its space, located in the buffer zone of the Protected Area (Table 2.5).

TABLE 2.5 Descriptive Analysis of Dimension Clarity of Message and Contribution to Conservation

Dimension Clarity of Message and Contribution to Conservation	Media	SD
20. Do you agree that there is a clear, relevant, articulated message between the activities and bird observatories that exist in the reserve?	4.00	1.076
21. Do you consider that the visit to the reserve generates a change in the positive behavior of visitors about biodiversity conservation?	4.65	0.587

In addition to the analysis of the dimensions, it is important to analyze the three main spaces linked to the birdwatching service offered by the reserve and the interpretation of the available heritage. The graph below shows the average rating obtained by each observatory considering all the dimensions mentioned above. The observatory with the highest rating is the hummingbird observatory (M=4.13), followed by the quail and partridge observatory (M=3.64) and tanagers (M=3.63) (Figure 2.3).

In this regard, as indicated above, the hummingbird observatory is the oldest, and a more significant number of species can be observed and seen for a longer period. Due to the characteristics of the species, birding in the Quail and Tinamou observatory is more complex, and in the case of the Tanager observatory, is the newest observatory and is only recently attracting species.

In addition, according to the on-site evaluation, which coincides with the results presented, in the three existing observatories, the information available in physical form is limited, and the existing research and monitoring are informal and require greater professionalization, which probably influenced

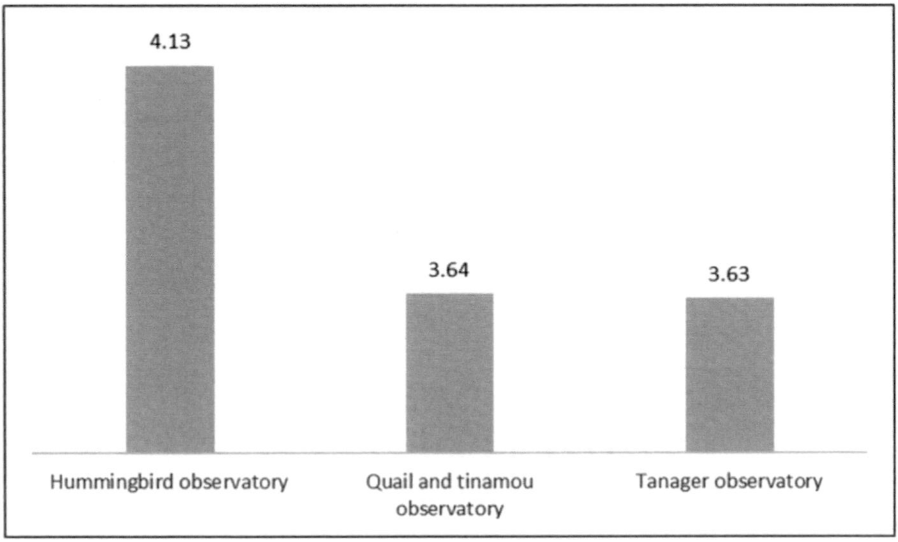

FIGURE 2.3 The average rating obtained by each observatory.

the results obtained. It is important to note that the manager's skills have been gained in the process of implementing birdwatching services in the reserve and that funding is limited, through support from non-governmental organizations and other conservation agencies.

In addition to the effectiveness in increasing the success of bird watching, bird observatories or bird hides contribute to reducing the intensity of disturbance and the level of bird response to human activity. According to Ma, Ng, Cheung, & Lam (2022), refuges or hides demonstrate a good level of capacity and effectiveness in mitigating the impacts of bird watchers, especially in terms of the frequency of bird response, the level of response they exhibit (alert response and prolonged flight), and associated visual disturbance behaviors.

2.6 CONCLUSION

From the evaluation, it can be concluded that the Arena Blanca Ecological Reserve has advanced in the process of heritage interpretation, and the main findings are as follows:

- The main activity performed is birdwatching, and birds are recognized as the main interpretive resources and attraction of the Reserve.

- An interpretive message is also recognized and contributes to visitor awareness of the importance of the site for biodiversity conservation, according to the objective of the Reserve and the relation to the Alto Mayo Protected Area.
- The lowest valued dimension is information and guidance, with the tanager observatory being the resource with the least information available. Currently, there is not enough physical information available; there is only a brochure with the 30 main species and the book "Birds of Peru," which is used as support. Information is provided by the guide who leads the group, complemented by the reserve manager, Norbil Becerra.
- The Reserve's facilities contribute to the interpretation of the heritage. In this case, the factor of support from these facilities to increase the success of several birds observed is the key to their valuation, which could increase if the information available in each facility were available and if the success of the birdwatching increases. Also, the observatories help to the success of birds watched, and at the same time reduce the intensity of disturbance and the level of bird response to human activity.
- For example, the least valued observatory was the quail and tinamou observatory, while the hummingbird observatory was the most valued, which coincides with the number of birds observed in each observatory. The quail and tinamou observatory is the least valued and the hummingbird observatory is the most valued.
- The research and monitoring dimension is also highly valued, which is key to ensuring the success of bird watching and providing adequate information on bird behavior. However, the limited formality and empirical development of monitoring, as well as the lack of systematization and creation of content, affect the dimension's valuation.
- Finally, it is very interesting to compare the three observatories, where the hummingbird observatory was the most highly valued, followed by the quail and tinamou observatory and the tanager observatory. In this regard, it can be inferred that the high valuation would be related to the dimensions evaluated that are related to the time of operation of the hummingbird observatory, the greater number of species available, and greater knowledge and information available from the manager, but could also be influenced by

the beauty and attractiveness of the species, the availability of more time to observe the species, among the most important.

Despite the lower qualification of the Quail and Tinamou observatory and the Tanager observatory, it is important to point out that these observatories are unique in the region since in sites that offer experience in bird watching, hummingbird observatories are relatively common, but not for other types of birds. In this sense, the Arena Blanca Ecological Reserve has a hummingbird observatory, which has some very outstanding species that are not seen in other areas, but the most relevant differentiation is for the implementation of the two additional observatories, which have species that are not easily seen throughout the region and help to increase the value of the birding site and the birding route.

2.7 RECOMMENDATIONS

Based on the results of the research conducted, it is recommended:

- Articulate an interpretation strategy that includes all the dimensions evaluated: Interpretive resources dimension, Research and monitoring dimension, Information and guidance dimension, Facilities dimension, and clarity of message and contribution to conservation dimension.
- Implement an improvement process for species research and monitoring that includes species recording, systematization of results, creation of publications or physical and virtual content, and elaboration of scientific articles.
- Implement fixed visual interpretive signals in the three bird observatories, based on the information available and the improvement process for species research and monitoring implemented. At the time of the field visit, the design of an interpretive poster for each observatory has been initiated to improve this dimension in the immediate future.
- Identify complementary interpretive resources for birds that can help reinforce the visitor experience and raise awareness of the importance of the Reserve for biodiversity conservation. The additional resources could be birds, other wildlife animals, agricultural products like coffee, or local rural activities.

KEYWORDS

- **birds**
- **birdwatching**
- **Heritage interpretation**
- **observatories**
- **Peru**
- **reserve**
- **tinamou observatory**
- **wildlife tourism**

REFERENCES

Apps, K., Dimmock, K., & Huveneers, C. (2016). Turning wildlife experiences into conservation action: Can White Shark cage diving tourism influence conservation behavior? *Marine Policy, 88*, 108–115.

Asociación para la interpretación del patrimonio (n.d). Principios teóricos de la interpretación. Interpretación del patrimonio. Retrieved on 18 June 2024 from https://interpretaciondelpatrimonio.com/interpretacion/

Brunelli, M. (2019). Monografía acerca de los orígenes de la interpretación del patrimonio en los Estados Unidos (1872–1920). Boletín de Interpretación, 40. Edición especial XX aniversario AIP. Asociación para la interpretación del Patrimonio. https://boletin.interpretaciondelpatrimonio.com/index.php/boletin/issue/view/43.

Chen, L. -J., & Chen, W. -P. (2015). Push – pull factors in international birders' travel. *Tourism Management, 48*, 416–425. https://doi.org/10.1016/j.tourman.2014.12.011

Connell, J. (2009). Birdwatching, twitching, and tourism: Towards an Australian perspective. *Australian Geographer, 40*(2), 203-217.

Dumbraveanu, D., Craciun, A., & Tudoricu, A. (2016). Principles of interpretation, tourism and heritage interpretation - The experience of Romanian museums. Human Geographies, 10(1), 59–75. https://doi.org/10.5719/hgeo.2016.101.4.

eBird. (n.d.). Listado de campo de eBird de Reserva Arena Blanca, San Martin, Peru. Retrieved on 18 June 2024 from https://ebird.org/printableList?regionCode=L3637689.

Hongyu, W., Beile, Z., & Hanqin, Q. (2022). How a hierarchical governance structure influencescultural heritage destination sustainability: A context of red tourism in China. *Journal of Hospitality and Tourism Management. 50*, 421–432, ISSN 1447-6770. https://doi.org/10.1016/j.jhtm.2022.02.002.

Ma, A. T. H., Ng, S. L., Cheung, L. T. O., & Lam, T. W. L. (2022). The effectiveness of bird hides in mitigating recreational disturbances of birdwatchers. *Journal for Nature Conservation, 67*, Article 126181. https://doi.org/10.1016/j.jnc.2022.126181.

Ma, Z., Cheng, Y., Wang, J., & Fu, X. (2013). The rapid development of birdwatching in mainland China: A new force for bird study and conservation. *Bird Conservation International, 23*(2), 259–269.

Mearns, K., & Botha, E. (2018). The Roles of Interpretation in the Management of National Parks in South Africa. *National Parks – Management and Conservation, 10*, 143–156.

Morales, J., & Ham, S. H. (2008). ¿A qué interpretación nos referimos? *Boletín de Interpretación, 19*, 4–7.

Orams, M. B. (2002). Feeding wildlife as a tourism attraction: A review of issues and impacts. *Tourism Management, 23*(3), 281–293.

Plenge, M. A. (2023). List of the birds of Peru / Lista de las aves del Perú. Boletín UNOP - Unión de Ornitólogos del Perú. Retrieved on 18 June 2024 from https://sites.google.com/site/boletinunop/checklist.

Promperu (2023a). The northern Peru birding route. Birding routes. Retrieved on 18 June 2024 from https://birdwatching.peru.travel/Imagenes/archivos/6/01-the-northern-peru-birding-route-2023.pdf

Promperu (2023b). The southern Peru birding route. Birding routes. Retrieved on 18 June 2024 from https://birdwatching.peru.travel/Imagenes/archivos/8/03-the-southern-peru-birding-route.pdf.

Promperu (2023c). The central Peru birding route. Birding routes. Retrieved on 18 June 2024 from https://birdwatching.peru.travel/Imagenes/archivos/7/02-the-central-peru-birding-route_compressed.pdf.

Puhakka L., Salo M., Sääksjärvi I.E. (2011). Bird Diversity, Birdwatching Tourism and Conservation in Peru: A Geographic Analysis. PLoS ONE 6(11): e26786. https://doi.org/10.1371/journal.pone.0026786.

Schwoerer, T., & Dawson, N. G. (2022). Small sight-Big might: Economic impact of bird tourism shows opportunities for rural communities and biodiversity conservation. PloS one, 17(7), e0268594. https://doi.org/10.1371/journal.pone.0268594.

Sekercioglu, C. H. (2002). Impacts of birdwatching on human and avian communities. *Environmental Conservation, 29*(3), 282–289.

Smithsonian's National Zoo & Conservation Biology Institute. (n.d.). Hummingbirds. Retrieved on 18 June 2024 from https://nationalzoo.si.edu/migratory-birds/hummingbirds.

Stemmer, K., Aas, O., Veisten, K., & Lindberg, K. (2022). Assessing recreation specialization to guide nature-based tourism development: A hybrid choice model of birder destination preferences. *Journal of Outdoor Recreation and Tourism, 39*, 1–12. https://doi.org/10.1016/j.jort.2022.100491.

U.S. Department of the Interior, U.S. Fish and Wildlife Service, and U.S. Department of Commerce, U.S. Census Bureau (2016). National Survey of Fishing, Hunting, and Wildlife-Associated Recreation.

World Tourism Organization (2014), Towards Measuring the Economic Value of Wildlife Watching Tourism in Africa – Briefing Paper, UNWTO, Madrid. https://doi.org/10.18111/9789284416752.

Young, B.E. (2007). Distribución de las especies endémicas en la vertiente oriental de los Andes en Perú y Bolivia. NatureServe, Arlington, Virginia, EE UU. https://sinia.minam.gob.pe/sites/default/files/siar-puno/archivos/public/docs/distribucion_de_las_especies_endemincas_en_la_vertiente_oriental_de_los_andes_de_peru_y_bolivia.pdf.

CHAPTER 3

CHALLENGES IN CONSERVATION OF ARCHITECTURAL HERITAGE AND TOURISM DEVELOPMENT IN URBAN AREAS: A CASE STUDY IN HO CHI MINH CITY, VIETNAM

NAM TRUNG NGUYEN,[1] BAO HA SON,[1] NGA VIET NGUYEN,[2] and THANH VAN VO[3]

[1]Researcher at Vietnam Tourism Research, Vietnam

[2]Faculty of Political Theory and General Knowledge, Ho Chi Minh City University of Culture, Vietnam

[3]Faculty of Tourism, Van Lang University, Ho Chi Minh City, Vietnam

ABSTRACT

This study explores the challenges of preserving and promoting architectural heritage values for tourism development in urban areas, with a focus on Ho Chi Minh City. Despite having a rich cultural heritage, particularly in French colonial architecture, the conservation and promotion of heritage face several challenges. The study uses qualitative methods to identify these challenges, including violation of the space and landscape of heritage, overexploitation of heritage, and inadequate heritage management. The Ho Chi Minh City authorities have prioritized the economic value of heritage, leading to messy and overloaded monuments. The overlapping functions and unclear responsibilities of stakeholders in heritage management also pose challenges. To effectively develop heritage tourism, close coordination among stakeholders

Dynamics of Heritage Management and Tourism Development: Trends, Practices, and Approaches. Surabhi Srivastava & Ruchika Kulshrestha (Eds.)

and balanced interests and responsibilities are necessary. However, implementing solutions remains a significant challenge. The study highlights the need for the state management agency to play an intermediary role and bring tourism in the right direction. The balance between preserving architectural heritage and developing tourism in urban areas is a long-standing challenge not only in Vietnam but also in other countries worldwide.

3.1 INTRODUCTION

Tourism has become a crucial driver of economic and cultural growth in urban areas, with architectural heritage being a popular resource for tourists (Tsiotsou & Ratten, 2010). However, balancing conservation efforts with tourism development poses significant challenges. This research focuses on Ho Chi Minh City, Vietnam, to identify and address obstacles in preserving and promoting architectural heritage values for tourism development.

Ho Chi Minh City has a rich cultural and architectural legacy, particularly in French colonial architecture, that offers potential for socioeconomic growth through tourism. However, there are challenges in conserving and promoting this legacy, including the violation of heritage space and landscape, overexploitation of heritage for commercial benefit, and insufficient heritage management (Nian et al., 2019). This study aims to identify these obstacles and suggest recommendations to enhance the value and potential of Ho Chi Minh City's cultural and architectural legacy for socioeconomic development.

The research employs qualitative tools, including observations, site surveys, interviews, and secondary data analysis on heritage, architecture, tourism, and conservation policy, to understand the current condition of conservation and tourist development in the city. The study offers two contributions: first, it outlines the present obstacles to the preservation of architectural history and the expansion of tourism in Ho Chi Minh City. Second, it offers suggestions for achieving a balance between conservation efforts and tourist growth, which may serve as a significant resource for lawmakers, tourism industry experts, and historical conservation groups in the city.

Cultural heritage tourism is a significant resource for tourism growth across the world, with many visitors choosing cultural heritage as their trip destination (Timothy, 2014). Cultural heritage tourism is expected to be worth around $556.96 billion worldwide in 2021 and is projected to increase between 2022 and 2030 (Research, 2023). Heritage tourism is a unique form of tourism characterized by the contradictory characteristics of uniqueness and universality. Unique qualities require distinct strategies for growing this form of tourism, while universal qualities necessitate a broader approach (Nuryanti, 1996). Strategies must also cater to different types of cultural tourists, such as "purposeful cultural tourists" and "accidental cultural tourists" (Ho & McKercher, 2004).

While the cohabitation of cultural heritage with tourism can bring economic benefits and foster local pride and cultural interchange (Kim & Ritchie, 2014; Ho & McKercher, 2004), excessive tourism can lead to the commodification and commercialization of cultural heritage, compromising its authenticity and integrity (Tsiotsou & Ratten, 2010). Tourism may also result in gentrification, displacing residents and transforming the social and cultural fabric of districts (Kim & Jamal, 2015).

To achieve sustainable and accountable cultural heritage tourism, urban planners, legislators, and tourism stakeholders must adopt effective regulation, community participation, and stakeholder collaboration (Wang & Gu, 2020). By striking a balance between the demands of tourist growth and the protection and promotion of cultural heritage, towns can capitalize on the promise of cultural heritage tourism while avoiding its potential negative effects. The purpose of this study is to advance knowledge and comprehension of the issues associated with protecting architectural heritage in urban environments and the potential of architectural heritage for sustainable tourist development.

3.2 ARCHITECTURAL HERITAGE CONSERVATION AND TOURISM DEVELOPMENT IN URBAN AREAS

The preservation and management of urban cultural sites require a comprehensive and collaborative approach involving various stakeholders such as local communities, government agencies, and commercial developers (Feyers, Stein, & Klizentyte, 2020). It is important to integrate heritage preservation into urban planning to support cultural preservation and sustainable development.

However, gentrification poses a severe threat to urban cultural buildings, especially in tourist-heavy towns. Therefore, conservation efforts must prioritize social justice and community engagement (González-Pérez, 2020). The rise of cultural and creative tourism has been a driving force behind economic growth worldwide, but it can have severe implications for local communities and cultural heritage (Richards, 2020). To achieve sustainable tourism development, various stakeholders need to be involved in striking a balance between economic benefits, cultural preservation, and environmental protection (Gao, Lin, & Zhang, 2021).

A multidisciplinary approach, including architectural design, engineering, cultural heritage administration, and urban planning, is necessary to protect urban architectural history (Karaman, 2022). The preservation and maintenance of architectural heritage should prioritize sustainability and long-term repair (Achmad & Yulianah, 2022).

While tourism can promote urban economic growth and heritage site promotion, it can also harm architectural heritage by increasing visitor numbers and commercializing culture. Hence, it is critical to strike a balance between historical preservation and promotion, with a focus on the local community and sustainable development (Santos-Júnior et al., 2020). Preservation and development require trade-offs between growth and preservation, taking into account the interests of the local population and the tourism sector, and requiring significant investment (Li et al., 2015).

In conclusion, sustainable management and preservation of architectural history in urban environments require a collaborative and multidisciplinary approach that balances preservation and development, takes into account the local community and tourism sector, and emphasizes long-term thinking and community participation in decision-making processes. Sustainable development entails creating economic opportunities that support historic preservation while also addressing social and environmental concerns.

3.3 THE CONTEXT OF HO CHI MINH CITY

The city of Ho Chi Minh, commonly known as Saigon, is a dynamic and culturally rich city in Southeast Asia with a population of almost 9 million people. Its architectural and cultural heritage reflects its rich history and influences. The city's architectural legacy is an essential part of its identity and one of its most popular tourist attractions. Colonial-era structures, such as the Saigon Central Post Office, designed by the French architect Gustave Eiffel, and the Notre-Dame Cathedral, are prominent examples of colonial architecture that offer a glimpse into the city's past. In contrast, the Giac Lam Pagoda, which dates back to the 18th century, exemplifies traditional Vietnamese architecture and is adorned with beautiful carvings.

Tourism development has played a crucial role in the city's economic prosperity and cultural preservation. With its rich cultural heritage and lively city life, Ho Chi Minh City has become a popular destination for travelers from around the world. The city is renowned for its street food scene, which features a diverse range of delicious and affordable foods that represent the city's various culinary traditions. The War Remnants Museum and the

Reunification Palace provide visitors with a deeper understanding of the city's history.

Ho Chi Minh City's thriving art and music scene, which includes galleries, theaters, and music venues, as well as the Ho Chi Minh City International Film Festival, contributes to its lively city life. The city's conservation policies and practices have successfully preserved its cultural history and unique identity. The restoration of Notre Dame Cathedral is an important conservation effort, and initiatives such as the "Green Lotus" program encourage sustainable tourism and ecologically friendly practices in the tourism sector.

Ho Chi Minh City is a fascinating tourist destination that offers visitors a glimpse into its rich cultural history, diverse cuisine, and unique blend of contemporary and traditional. Its architecture, museums, and landmarks reflect its history and cultural influences, providing tourists with an insight into its past. The city's dynamic environment, cultural opportunities, and active nightlife contribute to its appeal as a tourist destination. Through its preservation efforts, Ho Chi Minh City ensures that future generations can appreciate its distinctive character and cultural heritage (Figure 3.1).

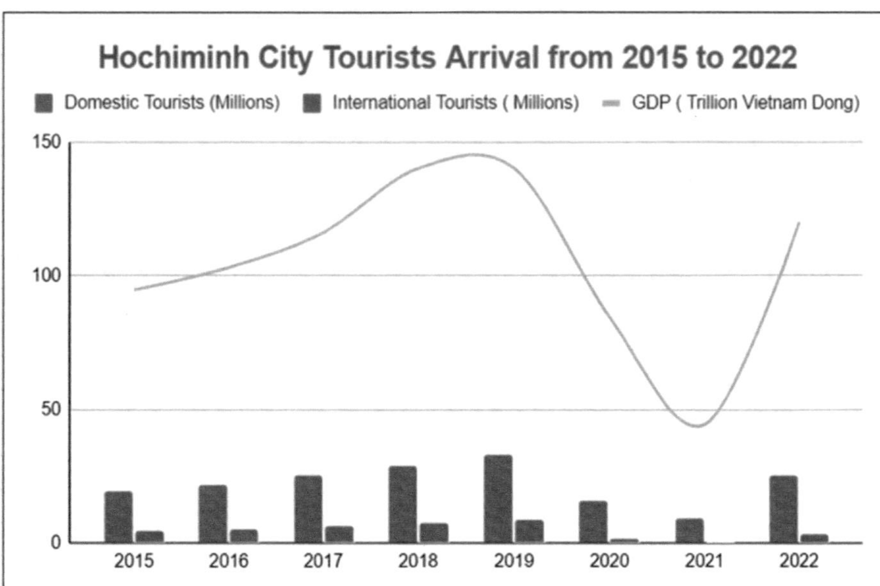

FIGURE 3.1 Tourists' arrival to Ho Chi Minh City from 2015 to 2022 (GSO, 2023).

The preservation and sustainable management of Ho Chi Minh City's architectural heritage is essential to maintain the city's identity, promote cultural tourism, and contribute to the local economy. The city has implemented several conservation policies and practices to protect its architectural heritage. For example, the restoration of the Notre Dame Cathedral is a significant conservation effort that ensures the preservation of the city's cultural legacy. Additionally, programs like the "Green Lotus" program encourage sustainable tourism practices that minimize the environmental impact of tourism.

Tourism has played a significant role in Ho Chi Minh City's economic growth and cultural preservation. The city's cultural attractions, such as its museums, landmarks, and street food scene, draw tourists from around the world. The city's art and music scene, which includes galleries, theaters, and music venues, adds to its vibrant city life. By promoting sustainable tourism practices, the city can attract visitors while preserving its cultural heritage and minimizing its impact on the environment.

In conclusion, Ho Chi Minh City's architectural heritage is an integral part of its cultural and historical identity. It serves as a reminder of the city's past, symbolizes its cultural influences, and contributes to its aesthetic appeal. The city's conservation policies and sustainable tourism practices are essential to preserve its architectural heritage, promote cultural tourism, and support its economic growth. By preserving and promoting its architectural heritage, Ho Chi Minh City can continue to attract visitors and maintain its unique and vibrant character (Table 3.1).

Ho Chi Minh City's architectural heritage is a vital component of the city's cultural and historical significance, attracting both domestic and international tourists to its recognizable structures and sites. The diverse and complex history of the city is reflected in the combination of Eastern and Western architectural styles found in its heritage buildings, such as the Notre Dame Cathedral, the Central Post Office, and the Reunification Palace, which have become iconic landmarks that draw tens of thousands of visitors each year.

In addition to its cultural and historical importance, the preservation and promotion of the city's architectural history also provide economic benefits. The maintenance and repair of these structures create job opportunities and contribute to the growth of tourism-related sectors, such as hotels and souvenir industries. The influx of tourists also generates revenue for local businesses and the community.

Maintaining and promoting Ho Chi Minh City's architectural heritage is crucial not only for the tourism industry but also for preserving the city's unique character and cultural identity for future generations. By continuing to celebrate and protect its architectural history, the city can ensure that its rich cultural legacy remains an important part of its growth and development.

TABLE 3.1 The Notable French Architectural Heritage in Ho Chi Minh City

Name	Year Built	Description
Saigon Notre-Dame Basilica	1863–1880	A Catholic cathedral located in the downtown area, built with red bricks imported from Marseille, France.
Central Post Office	1886–1891	A landmark building designed by Gustave Eiffel, featuring neoclassical and Gothic architectural styles.
Municipal Theatre	1897	Also known as the Saigon Opera House, it was designed by French architect Eugène Ferret and serves as a venue for cultural performances.
Ho Chi Minh City Museum of Fine Arts	1929	A colonial-era building designed by architect Rivera now displays contemporary Vietnamese art.
Hotel Majestic Saigon	1925	A luxury hotel built in the colonial style, featuring Art Deco elements.
Independence Palace	1966	Formerly known as the Presidential Palace, it was the home of the South Vietnamese president during the Vietnam War and now serves as a museum.
Binh Tay Market	1928	A bustling wholesale market located in the city's Chinatown district, featuring a mix of Chinese and French architectural elements.
The Continental Saigon	1880	A historic hotel located in the heart of the city, featuring a colonial-style façade and luxury accommodations.

Note that this is not an exhaustive list and there are many other French architectural heritage sites in Ho Chi Minh City.

3.4 CHALLENGES IN PRESERVING ARCHITECTURAL HERITAGE AND DEVELOPING TOURISM IN HO CHI MINH CITY

Ho Chi Minh City (HCMC) has a rich cultural and architectural legacy, but its preservation faces a number of obstacles. This section will explore the violation of architectural historic space and landscape, the overexploitation of heritage, and the insufficient management of heritage.

3.4.1 THE LACK OF EFFECTIVE PLANNING AND RECOGNITION OF CULTURAL HERITAGE

The growth of heritage tourism in Ho Chi Minh City is hindered by the absence of adequate planning and protection for cultural assets. Despite the city's rich history and cultural heritage, many ancient architectural structures are used for other purposes and not encouraged for tourist visits, resulting in missed opportunities to attract tourists and stimulate the tourism sector's economic growth.

The State Bank of Vietnam and the Archbishop's Palace are not available to the public for viewing, while other old villas that could be transformed into museums or other cultural places for visitors are rented out as restaurants and cafés. This situation has slowed the growth of heritage tourism in Ho Chi Minh City, reducing the number of opportunities to explore historical monuments and placing these structures at risk of being demolished or damaged.

Furthermore, a lack of financing for preservation activities exacerbates the absence of acknowledgment and protection for these ancient sites. Many of these old structures require significant investment to maintain and protect them for future generations. Still, heritage conservation is rarely given the same importance as other industries such as infrastructure development. The city has divided its heritage-worthy urban architectural works into two categories based on current rules and management procedures. Group 1 consists of relics maintained by the Cultural Heritage Law of 2002, while Group 2 consists of architectural and landscape areas of conservation importance that are not relics. HCMC faces challenges in managing works belonging to both groups, including the development of new structures within relic sites that detract from their cultural value and the management of construction in Group 2 regions in accordance with the Heritage Law.

According to a 2017 study and evaluation of tourist resources undertaken by the HCMC Tourism Department in partnership with Hue University, relics, and ancient architecture are directly contributing to the growth of the HCMC tourism business. However, there are just 258 tourism-related resources in the city, and only 111 of these are viable. Approximately 30% of the city's 172 treasures may be exploited as tourist attractions. Nonetheless, according to foreign travel websites, all ten must-see places in HCMC are historic architecture.

Many French architectural achievements remain in HCMC, while the Chinese have left their mark on the Binh Tay market, temples, and pagodas. They are both great assets and an issue for legacy cities, as locals become concerned about the ownership of recognized heritage works. Several residents of ancient villas are currently in a difficult situation, as their land is highly valuable but cannot be sold, transferred, or rebuilt. The dilemma is how to effectively protect history without provoking disputes and compensating those who gain from development. Presently, eighteen artifacts and old architectural structures have been destroyed. The protected historical region was overlooked during urban transit planning, and the classification of relics was sluggish, resulting in numerous relics being situated in planned areas. Due to urbanization, population increase, and pressure on land finances to invest in economic and social development projects, cultural artifacts and urban architectural spaces have been distorted, vanished, or are at risk.

The preservation of cultural heritage is crucial for maintaining the identity and cultural legacy of a city. However, legislation for historical preservation is still lagging behind the pace of economic and social growth. Additionally, inadequate execution of these regulations and a lack of resources for conservation efforts are contributing factors to the issue. Numerous studies have addressed the difficulties of balancing urban tourist development and the preservation of cultural assets.

For instance, studies conducted in China by Zhuang, Yao, & Li (2019) and Weng et al. (2019) found that rapid tourism development negatively affected the preservation of heritage, resulting in the destruction of historical buildings and the loss of cultural identity. Similarly, conflicts between tourism development and heritage conservation in urban areas of Korea were identified by Kim & Jamal (2015). They suggested a comprehensive approach that includes stakeholder participation, efficient management, and the development of sustainable tourism to achieve a balance between tourism development and heritage conservation. In the Iranian city of Kashan, Eshrati et al. (2017) reported similar findings and recommended a sustainable tourism strategy that incorporates heritage site preservation and stakeholder participation in the decision-making process.

The difficulties of balancing tourism growth with the preservation of cultural heritage in Ho Chi Minh City, Vietnam, are comparable to those described in earlier studies. The rapid growth of tourism in the city has put pressure on the preservation of cultural assets, leading to the repurposing of historic structures and the loss of cultural identity. However, comprehensive solutions that incorporate stakeholder engagement, efficient management, and the development of sustainable tourism are lacking (Nguyen, 2021). Therefore, it is crucial for the government and relevant stakeholders to address these issues and work together towards the preservation and promotion of cultural heritage in Ho Chi Minh City.

3.4.2 THE DISTURBANCE OF ARCHITECTURALLY SIGNIFICANT LANDSCAPE AND SPACE

The violation of HCMC's architectural heritage space and landscape is a complicated and multidimensional problem. Contributing issues include rapid urbanization and development, the encroachment of new developments on historic structures, and poor heritage management and planning. These difficulties are not specific to Ho Chi Minh City but are shared by cities throughout the globe. In the previous decade alone, nearly 10,000 cultural and historical heritage structures have been lost in HCMC (HCMC People's Committee 2020). For instance, according to our study, the loss of French colonial architectural history in Ho Chi Minh City, notably on Nguyen Hue Street and in District 3, is a significant issue. Many new office buildings of diverse architectural types have replaced old structures, signifying a transition towards a more contemporary urban setting. Consideration should also be given to the elimination of renowned heritage places such as Tax Market, Hoang De Hotel, and Ba Son Shipyard, which played a vital role in Vietnam's shipbuilding sector. Despite the efforts of experts and residents to conserve these cultural and historical treasures, many have been lost. The findings illustrate the difficulties associated with protecting architectural history in a fast-expanding metropolis, where economic expansion and urbanization sometimes take precedence over conservation initiatives. It also refers to the loss of cultural and historical identity due to urbanization and modernization.

The destruction of ancient districts and structures in Beijing, China (Zhuang, Yao, & Li 2019) and the overexploitation of historic sites in Istanbul, Turkey (Burak, Doğan, & Gazioğlu 2004) are examples of issues encountered by other cities. Moreover, tourism can contribute to the destruction of architectural heritage spaces and landscapes. In Venice, Italy, the rush of visitors has caused congestion and damage to historic structures and public spaces (Bertocchi et al., 2020), while in Petra, Jordan, misuse of paths and deterioration of the natural landscape have happened (Knodell et al., 2017). Due to the increasing rise of tourism in the city, HCMC is also facing comparable issues.

The destruction of architecturally significant landscapes and space is a global issue that affects cities all over the world. As cities expand and develop, historic structures and public spaces are often neglected or destroyed in the process. The loss of these cultural and historical treasures can have a profound impact on the identity and character of a city.

One of the main drivers of the destruction of architectural heritage spaces and landscapes is rapid urbanization and development. This is particularly true in rapidly developing countries like Vietnam, where economic expansion and urbanization are prioritized over conservation efforts. In Ho Chi Minh City, nearly 10,000 cultural and historical heritage structures have been lost in the past decade alone, which highlights the scale of the problem (HCMC People's Committee, 2020).

Another contributing factor is poor heritage management and planning. Without effective planning and management strategies in place, heritage structures and public spaces are often neglected or overlooked. This is compounded by the encroachment of new developments on historic structures, as developers prioritize profit over conservation.

Tourism can also contribute to the destruction of architectural heritage spaces and landscapes. As cities become more popular tourist destinations, the influx of visitors can cause congestion and damage to historic structures and public spaces. This is evident in cities like Venice, where the sheer number of visitors has caused damage to historic structures and public spaces.

In conclusion, the destruction of architecturally significant landscapes and spaces is a complex and multidimensional problem that affects cities all over the world. Efforts to conserve heritage structures and public spaces must be prioritized by city officials and developers, and effective planning and management strategies must be put in place to ensure the preservation of these cultural and historical treasures. The balance between economic development and heritage conservation must be struck, and stakeholders must be engaged in the decision-making process to ensure that heritage conservation is given the attention it deserves.

3.4.3 INADEQUATE HERITAGE MANAGEMENT AND PLANNING FOR PROMOTING HERITAGE SITES

The degradation of HCMC's architectural heritage landscape and space is a complex and multifaceted issue, influenced by rapid urbanization, new developments encroaching on historic structures, and inadequate heritage management and planning. Unfortunately, these challenges are not unique to Ho Chi Minh City but are encountered by cities worldwide. Shockingly, nearly 10,000 cultural and historical heritage structures have been lost in HCMC alone in the last decade (HCMC People's Committee 2020). Our research highlights the significant loss of French colonial architectural history in HCMC, particularly on Nguyen Hue Street and in District 3, due to the replacement of old buildings with new office towers. Renowned heritage sites such as Tax Market, Hoang De Hotel, and Ba Son Shipyard have also been eliminated, despite efforts by experts and residents to conserve them. The findings underscore the difficulties of preserving architectural heritage in a rapidly expanding metropolis, where economic growth and urbanization often take precedence over conservation initiatives. This destruction also erodes cultural and historical identity due to modernization and urbanization.

Beijing, China's demolition of ancient districts and structures (Zhuang, Yao, & Li, 2019) and Istanbul, Turkey's overexploitation of historic sites (Burak, Doğan, & Gazioğlu, 2004) are examples of similar issues faced by other cities worldwide. Furthermore, tourism can contribute to the destruction of architectural heritage spaces and landscapes. In Venice, Italy, the overwhelming influx of tourists has caused congestion and damage to historic structures and public spaces (Bertocchi et al., 2020), while in Petra, Jordan, misuse of paths and deterioration of the natural landscape have occurred (Knodell et al., 2017). HCMC is similarly confronting similar issues due to the growing tourism industry in the city.

Indeed, the lack of thorough research and assessments of the system of objects that must be protected in terms of architectural landscapes poses a significant challenge in the formulation of specific guidelines and laws for the conservation of architectural landscapes. Without a clear understanding of the objects that need to be protected, it becomes difficult to develop effective conservation strategies.

Furthermore, the lack of clarity on the duties and responsibilities of different sectors also hampers conservation efforts. In order to effectively conserve architectural landscapes, it is important for all stakeholders to understand their roles and responsibilities. This includes government agencies responsible for heritage conservation, as well as private developers and other stakeholders who may have an impact on architectural landscapes.

Improving cultural heritage laws and regulations is a crucial step towards better conservation of architectural landscapes. These laws must be comprehensive and up-to-date to address the specific challenges faced by different cities and regions. The lack of coordination between cultural institutions, as acknowledged by the Department of Planning and Architecture, also needs to be addressed. Stronger connections between museums, cultural houses, and heritage sites can improve the preservation and promotion of cultural heritage values.

In summary, addressing the challenges faced in the conservation of architectural landscapes requires a multi-faceted approach that includes comprehensive research and assessments, clear delineation of roles and responsibilities, and effective laws and regulations. Stronger coordination between cultural institutions is also necessary to ensure that cultural heritage values are preserved and promoted.

3.4.4 LACK OF EFFORTS TO PROMOTE CULTURAL HERITAGE SITES AND MAKE THEM COME ALIVE FOR TOURISTS

In the development of heritage tourism, it is critical to make cultural heritage sites in Ho Chi Minh City more engaging for visitors and immerse them in the daily lives of locals. While some museums and cultural sites, like Independence

Palace, have implemented various forms of interpretation, others have not made similar efforts to engage tourists. This lack of interaction may diminish the appeal of historical tourism for travelers seeking immersive experiences that allow them to connect with local communities and learn about their cultures.

The absence of efforts to promote and bring cultural heritage sites to life for visitors can have significant negative impacts on both the sites themselves and the communities surrounding them. Without effective promotion and interpretation, these sites may fail to attract visitors, which can lead to a lack of funding for their upkeep and preservation. This, in turn, can cause the sites to deteriorate and lose their value as cultural and historical resources.

Furthermore, when cultural heritage sites are not brought to life for visitors, the experiences of those who do visit may be shallow and unsatisfying. Visitors may not fully appreciate the historical or cultural significance of the site, which can limit their understanding and appreciation of the local community and its traditions. This can lead to missed opportunities for cultural exchange and understanding, and can also result in a lack of interest in preserving these sites for future generations.

Moreover, there is a significant problem with the preservation and marketing of privately held palaces, such as 138 Chau Van Liem Street, District 5; 41 Tu Xuong Street (Ward 7, District 3); 251 Dien Bien Phu Street (Ward 7, District 3); and 9 Ngo Thoi Nhiem Street (Ward 6, District 3), which hold historical and cultural value as they depict the lifestyles and riches of notable families from the past. However, due to budgetary constraints and lack of government assistance, these palaces often fall into disrepair and are inaccessible to the general public. Without adequate preservation and promotion, these palaces threaten to fade into oblivion, along with a portion of the city's history and cultural legacy.

The government and private individuals or groups should collaborate to conserve and promote these palaces as part of the city's cultural heritage. This may include funding for repairs and upkeep, as well as developing opportunities for the public to visit these palaces and learn about their history. In addition to preserving and promoting the city's cultural heritage, this would create opportunities for the local economy to benefit from historical tourism.

It is crucial to involve local people in the development and management of cultural sites to make heritage tourism more sustainable and inclusive. This can help ensure that the presentation of cultural heritage is accurate and respectful and that locals benefit from tourism growth.

In summary, the conflict between the requirements of urban development and the preservation of architectural landscapes has not been satisfactorily resolved. The conservation of architectural landscapes has been overshadowed by the need to address the economic expansion of urban development. It is vital to reevaluate the method of preserving old villas and categorize and differentiate them based on their historical uniqueness, architectural style, or association with significant historical events. In terms of architectural landscapes, it is essential to conduct comprehensive and systematic research and assessments of the items requiring conservation, establish specific guidelines and criteria, and identify the roles and responsibilities of various sectors and levels in the preservation of architectural landscapes. There is a need to improve cultural heritage laws and regulations to raise awareness and foster better preservation and promotion of architectural landscapes.

3.5 STRATEGIES TO BALANCE CONSERVATION AND TOURISM DEVELOPMENT

Ho Chi Minh City is a popular tourist destination due to its rich cultural legacy, including historical sites, architecture, and cultural customs. However, the increase in tourism has put a strain on the city's efforts to preserve its past. To achieve a balance between conservation and tourist growth, several measures can be taken.

Firstly, improving stakeholder collaboration and coordination is crucial. This involves government agencies, commercial firms, non-governmental groups, and local communities working together to build a clear vision and plan for cultural preservation and tourist development. This can be achieved through frequent meetings, cooperative planning sessions, and open communication to resolve problems and discover opportunities for cooperation. Setting standards for responsible tourism practices that protect the city's cultural sites and traditions can also be a part of this collaboration.

Secondly, increasing public understanding and engagement in cultural resource conservation is essential. Public education on the significance of historical preservation and the influence of tourism on the cultural character of the city is crucial. Educational initiatives, public events, and social media can be used to educate the public, with the municipal administration sponsoring cultural events and heritage preservation groups giving walking tours highlighting the architectural history of the city.

Thirdly, creating sustainable tourism products that protect and promote architectural heritage is necessary. The city can collaborate with tourist enterprises to establish sustainable tourism goods that protect the city's historic landmarks for future generations. Ecotourism goods that promote environmentally responsible practices, cultural tours that highlight the city's historical sites, and responsible tourism practices can be promoted.

Moreover, strategies for promoting and interpreting cultural heritage sites can be developed, including the use of digital technologies such as virtual reality and augmented reality. Partnering with local communities to develop cultural programming and events that showcase the significance and value of these sites is another approach. Providing interactive cultural heritage experiences, such as living history experiences and seminars, classes, or demonstrations of traditional arts or activities, can also help tourists better understand the daily lives, customs, and traditions of local populations (Table 3.2).

TABLE 3.2 Solution Frameworks to the Balance Conservation and Tourism Development

Solution Frameworks	Description
Improving stakeholder collaboration and coordination.	Collaborating among government agencies, commercial firms, non-governmental groups, and local communities to build a clear vision and plan for cultural preservation and tourist development.
Increasing public understanding and engagement in cultural resource conservation.	Educating the public on the significance of historical preservation and the influence of tourism on the cultural character of the city through different avenues of communication, including educational initiatives, public events, and social media.
Creating environmentally responsible tourist goods that protect and promote architectural heritage.	Developing sustainable tourism goods that protect and promote architectural heritage, such as ecotourism goods and cultural tours that highlight the city's historical sites and encourage responsible tourist practices.
Promoting and interpreting cultural heritage sites.	Using digital technologies such as virtual reality and augmented reality to create immersive and engaging experiences for visitors, partnering with local communities to develop cultural programming and events that showcase the significance and value of these sites, and offering living history experiences and demonstrations of traditional arts or activities.

By implementing these measures, Ho Chi Minh City can achieve its tourism development objectives while conserving its cultural character for future generations.

In conclusion, a multi-stakeholder approach is necessary to achieve a balance between conservation and tourism development in Ho Chi Minh City. This includes strengthening collaboration and coordination between stakeholders, increasing public awareness and participation in heritage conservation, and developing sustainable tourism products that preserve and promote architectural heritage. By applying these measures, the city can achieve its tourism development objectives while conserving its cultural character for future generations.

3.6 CONCLUSION

The preservation and promotion of Ho Chi Minh City's architectural heritage is vital for its socioeconomic development, cultural identity, and communal pride. However, this goal faces several challenges that require coordinated efforts among stakeholders, increased public awareness and participation, and the creation of sustainable tourism products. These efforts can produce job opportunities and help expand the tourism sector while preserving the city's cultural character.

Future studies might explore the possibilities of new technologies like virtual reality and augmented reality in promoting architectural heritage for tourism growth. In addition, detailed research is required to evaluate the efficacy of present conservation initiatives and propose areas for improvement.

The protection and promotion of architectural heritage for tourism development in Ho Chi Minh City is a difficult subject requiring a multidimensional strategy. It is feasible to achieve a balance between conservation and tourist development for the benefit of the local community and future generations by boosting cooperation between stakeholders, increasing public knowledge and engagement, and generating sustainable tourism products.

KEYWORDS

- **art architecture**
- **cultural heritage**
- **Heritage conservation**
- **Ho Chi Minh City**
- **tourism development**
- **urban tourism**
- **Vietnam**

REFERENCES

Achmad, W., & Yulianah, Y. (2022). Corporate Social Responsibility of the Hospitality Industry in Realizing Sustainable Tourism Development. *Enrichment: Journal of Management, 12*(2), 1610–1616.

Bertocchi, D., Camatti, N., Giove, S., & van der Borg, J. (2020). Venice and Overtourism: Simulating Sustainable Development Scenarios through a Tourism Carrying Capacity Model. *Sustainability*, *12*(2), 512. https://doi.org/10.3390/su12020763.

Burak, S., Doğan, E., & Gazioğlu, C. (2004). Impact of Urbanization and Tourism on Coastal Environment. *Ocean and Coastal Management, 47*(9–10), 515–527.

Eshrati, P., Fadaei Nezhad Bahramjerdi, S., Eftekhari Mahabadi, S., & Azad, M. (2017). Evaluation of authenticity on the basis of the Nara grid in adaptive reuse of manochehri historical house Kashan, Iran. *Archnet-IJAR, 11*(3), 214–230.

Feyers, S., Stein, T., & Klizentyte, K. (2020). Bridging worlds: Utilizing a multi-stakeholder framework to create extension-tourism partnerships. *Sustainability, 12*(1), 1–23.

Gao, J., Lin, H., & Zhang, C. (2021). Locally situated rights and the 'Doing' of responsibility for heritage conservation and Tourism Development at the Cultural Landscape of Honghe Hani Rice Terraces, China. *Journal of Sustainable Tourism, 29*(2–3), 193–213. https://doi.org/10.1080/09669582.2020.1727912.

González-Pérez, J. M. (2020). The dispute over tourist cities. Tourism gentrification in the Historic Centre of Palma (Majorca, Spain). *Tourism Geographies, 22*(1), 171–191. https://doi.org/10.1080/14616688.2019.1586986.

HCMC People's Committee. (2020). Preserving heritage and urban architectural landscape in the city, https://vpub.hochiminhcity.gov.vn/portal/pages/2019-11-20/Bao-ton-di-san-va-canh-quan-kien-truc-do-thi-tren-wtro9hg7bey6.aspx.

Ho, P. S. Y., & McKercher, B. (2004). Managing heritage resources as tourism products. *Asia Pacific Journal of Tourism Research, 9*(3), 255–266.

Karaman, Z. T. (2022). An understanding of inter and transdisciplinary aspects of Urban Resilience. *Open Journal of Social Sciences, 10*(04), 195–215.

Kim, S., & Jamal, T. (2015). The co-evolution of Rural Tourism and Sustainable Rural Development in Hongdong, Korea: Complexity, conflict and local response. *Journal of Sustainable Tourism, 23*(8–9), 1363–1385.

Knodell, A. R., et al. (2017). The Brown University Petra Archaeological Project: Landscape Archaeology in the Northern Hinterland of Petra, Jordan. *American Journal of Archaeology, 121*(4), 621–683.

Li, Y., Hunter, C., Hassan, A., & Rahman, M. (2015). Community Involvement for Sustainable Heritage Tourism: A Conceptual Model. *Journal of Cultural Heritage Management and Sustainable Development, 5*(3), 248–262.

Nguyen, T. H. (2021). Current Status of Urban Architectural Landscape Conservation in Ho Chi Minh City. Retrieved from: https://fudozon.com/tin-tuc/thuc-trang-cong-tac-bao-ton-canh-quan- kien-truc-do-thi-tren-dia-ban-thanh-pho-ho-chi-minh--3753.html (accessed on 18 June 2024).

Nian, S., et al. (2019). How Outstanding Universal Value, Service Quality and Place Attachment Influences Tourist Intention Towards World Heritage Conservation: A Case Study of Mount Sanqingshan National Park, China. *Sustainability, 11*(12), 1–18.

Nuryanti, W. (1996). Heritage and Postmodern Tourism.

Research, Grand view. (2023). Heritage Tourism market size, share & trends analysis report by type (Cultural, Natural, Intangible), by end-user (30 & under, 31 To 50, 51 To 70, 71 And above), By Booking Channel, By Region, And Segment Forecasts, 2022 – 2030.

Richards, G. (2020). Designing Creative Places: The Role of Creative Tourism. *Annals of Tourism Research, 85*(March), 102922. https://doi.org/10.1016/j.annals.2020.102922.

Santos-Júnior, A., Almeida-García, F., Morgado, P., & Mendes-Filho, L. (2020). Residents' quality of life in smart tourism destinations: A theoretical approach. *Sustainability, 12*(20), 1–24.

Timothy, D. J. (2014). Contemporary cultural heritage and tourism: Development issues and emerging trends. *Public Archaeology, 13*(1–3), 30–47.

Tsiotsou, R., & Ratten, V. (2010). Future research directions in tourism marketing. *Marketing Intelligence and Planning, 28*(4), 533–544.

Wang, S., & Gu, K. (2020). Pingyao: The historic urban landscape and planning for heritage-led urban changes. *Cities, 97*, 102489.

Zhuang, X., Yao, Y., & Li, J. (2019). Sociocultural Impacts of tourism on residents of world cultural heritage sites in China. *Sustainability, 11*(3), 840.

CHAPTER 4

DIGITIZATION AS A TOOL FOR CONSERVATION AND SUSTAINABILITY OF THE BUILT HERITAGES: A DISTANT DREAM OR AN EXISTING REALITY?

JAVED SALIM,[1] ANUBHA,[1] SARAH HUSSAIN,[2] NUSRAT YASMEEN,[2] and PRIYAKRUSHNA MOHANTY[3]

[1]*Research Scholar, Department of Tourism and Hospitality Management, Jamia Millia Islamia (A Central University), New Delhi, India*

[2]*Department of Tourism and Hospitality Management, Jamia Millia Islamia (A Central University), New Delhi, India*

[3]*Assistant Professor, Department of Business and Management, Christ University, Bangalore, Karnataka, India*

ABSTRACT

The idea of built heritage is as old as human civilization, and its objective stretches far beyond our understanding of the past. These built heritages as artifacts express humanity's history, diverse cultures, skills, and experiences exchanged and shared across generations. They provide specific characteristics for places, making them the material manifestation of social, economic, political, territorial, and environmental values. Unfortunately, natural and man-made disasters have been posing repetitive threats to the sustainability

Dynamics of Heritage Management and Tourism Development: Trends, Practices, and Approaches.
Surabhi Srivastava & Ruchika Kulshrestha (Eds.)

of these built heritages. Therefore, there has been a growing interest in and calls for the conservation and protection of these built heritages.

As we have ushered in the era of technical advancements, the world is abuzz with the word "digitization." The continual adaptation of digital technologies in various spheres of life and the growing popularity of digital devices have revolutionized consumption patterns in recent times. Consequently, the use of digital technologies to conserve and promote monuments has gained prevalence in various parts of the West. Further, in recent times, this trend seems to be catching on in developing countries like India as well.

However, given the novelty of the concept, the adoption and acceptance of the concept of digitized monuments are still surrounded by many unanswered questions. The study aims to reveal the opinions of the tourists visiting the various built heritage sites and understand their experiences with digital heritage and its influence on the tourist experience while visiting a built heritage monument. The chapter also tries to dive deep into their thoughts and understanding of the influence of digitization in promoting sustainable consumption.

In an attempt to achieve the aforementioned objective of the study, the data would be collected from the urban populace of India. A qualitative approach has been adopted to conduct the study, and semi-structured interview forms consisting of open-ended questions have been used to collect the data. MAXQDA software would be employed to analyze the responses collected and draw conclusions from the data collected. The study intends to bring out the opinion of tourists in regard to the digitization of built heritage in India. As the trend of digitization of built heritage is still in its nascent stage in India, only a few heritage monuments have been digitized in recent times.

4.1 INTRODUCTION

Heritage shapes part of the history, traditions, and cultural values of each society or community and is significant because it substantially impacts our sense of self, our loyalty, our knowledge, our beliefs, and our actions (Borissova, 2018; Gantait, Mohanty, & Swamy, 2018). Not only are monuments, materials, structures, and sites of historical, artistic, archaeological, scientific, or anthropological significance a part of cultural heritage, but it also includes traditions or living expressions inherited from ancestors and passed on to future generations and consists primarily of folklore, performing arts, social customs, rites, and festivities; knowledge and awareness concerning nature and the universe; or the skills and expertise necessary to produce traditional crafts (UNESCO, n.d.). The worldwide scenario of turmoil arising from a never-ending spell of wars, instability, terrorism, natural calamities, etc., has severely impacted and destroyed several irreplaceable historical pieces of the early human settlements (Khan, Shafi, & Ahangar, 2018; Manžuch, 2017). Heritage has a tremendous influence on the economic and social development of an area by contributing to tourism and giving a regional identity to give the area a competitive advantage over others (Ratrrout, Zidan, Abulhouf, & Sabha, 2006). Hence, globally, the cultural legacy should be protected for economic, cultural, and social reasons (Borissova, 2018). In addition to conventional approaches for the preservation and protection of built heritage, digitization is an emerging technology that may make substantial contributions in this regard (Pavlidis et al., 2007).

It is vital to keep the built heritage we already have from falling apart or being lost. According to UNESCO's "Fundamental Principles of Documentary Heritage Digitization," digitization is the process of converting physical, analog originals into digital files using a scanner, camera, or other electronic equipment. It is part of a process that comprises selection, evaluation, inclusion of requirements, prioritization, preparation of originals for digitization, collecting and production of metadata, digitization, and construction of data collections, and submission of digital resources to delivery systems and archives. This procedure is supported by management, including administration of intellectual property rights and quality control, and assessment at its conclusion (Bountouri, 2017). UNESCO (2003) defines "digital heritage" as "made up of computer-based materials of enduring value that should be kept for future generations. There are various conventional techniques to digitize a material of historical significance, but until recently, built heritage has been preserved through 3D models (Fiorillo et al., 2013; Pavlidis et al., 2007), which is much more appealing and attracts a larger audience. Digitization allows the preservation of authentic historical relics under ideal conditions and creates a digital version that is accessible to the public at large, which enlarges the business potential of activities such as tourism. Hence, preservation, accessibility, and promotion are some of the advantages of digitization (Balandin & Laizane, 2013; Chowdhury, 2002; Bachi, Fresa, Pierotti, & Prandoni, 2014).

One could trace the emergence of new technologies, i.e., digital recording or digitization of cultural heritage, despite the traditional conservation methods in the early 2000s in the West (Khan, Shafi, & Ahangar, 2018;

Ognjanović, Marinković, Šegan-Radonjić, & Masliković, 2019). In India, digitization through 3D models is a recent phenomenon (Singh, 2017), though there are organizations under state authorities limited to digitally documenting materials related to culture and heritage through conventional techniques (Ahmad & Sharma, 2020). The increasing availability of very inexpensive and simple-to-learn digitization technologies, along with the growth of Web 2.0, has fostered amateur digitization and the formation of non-professional digital archives managed by people and groups (Manžuch, 2017). This 3D VR (virtual reality) and AR (augmented reality) experience through the API (Application Programming Interface) encourages the masses to engage with it, but digitization poses few challenges (Mohanty, Hassan, & Ekiz, 2020). It is essential to keep in mind that the protection of heritage through digitization should not interfere with or even replace the protection of the built heritage but instead promote a new method for public dissemination and the efficient and appropriate commercialization of cultural resources (He, Ma, & Zhang, 2017).

This study aims to understand the dynamics behind the adoption and acceptance of the concept of digitized monuments. This study further reveals the opinions of the tourists visiting the various built heritage sites and understands their experiences with digital heritage and its influence on the tourist experience while visiting a built heritage monument. The chapter also tries to dive deep into tourists' thoughts and understanding of the influence of digitization in promoting sustainable consumption of built heritage.

4.2 LITERATURE REVIEW

One of the oldest skills to emerge in early civilizations was constructing different sorts of structures to meet basic human requirements (e.g., worship, refuge, shelter). The majority of the first human settlements were rural. Rural civilizations changed and became urban ones when an excess of production started to arise and cities started to develop. Buildings' complexity and variety have grown throughout time. At the same time, several facets of life and society were tied to and reflected in architecture. Human history and various civilizations are reflected via monuments, structures, and urban layouts. The architectural legacy, also known as built heritage, is one of the most significant parts of cultural heritage and is of exceptional importance; consequently, it must be conserved as a part of humanity's global history (Pavlidis et al., 2007).

Earlier conservation practices include traditional methods like microfilming, microfiche preparation, and storage in a controlled environment with the necessary temperature and humidity, etc. (Bakshi, 2016). In addition to conventional ways, modern technologies may greatly contribute to the preservation and display of architectural buildings. Recording monuments, buildings, locations, and cities digitally may also contribute to the preservation, exhibition, and transmission of architectural heritage. Nowadays, the growth and development of new recording and visualization methods enable the creation of 3D models based on reality that have a high degree of geometric precision and a realistic look. The resulting 3D products can then be used for detailed and accurate documentation and digital preservation of existing tangible heritages, as well as for new archaeological/architectural studies and analyses, 3D repositories and catalogs, virtual reconstruction, computer-aided restoration, physical replicas, VR and AR applications, multimedia museum exhibitions, and virtual visits (Remondino, Barazzetti, Nex, Scaioni, & Sarazzi, 2011). According to He, Ma, & Zhang (2017), with the development of photogrammetry and computer visualization technology, the digitization of cultural assets has progressed from 2D to 3D and even 4D. Many overlapping and interconnected factors make the distinction between these worlds ambiguous. Research by Fiorillo et al. (2013) compares 3D models with historical drawings of monuments in Italy and demonstrates coherence may be used to determine the precision of building heritage digitization.

Heritage is a resource for inter-institutional and cross-sector collaboration in economic, social, and culturally sustainable growth. Through its promotion and dissemination, culture and heritage will be shown to have actual social, economic, and cultural significance (Khan, Shafi, & Ahangar, 2018). The dissemination of information about the world's cultural heritage will assist individuals of other ethnicities to better understand one another (Mannoni, 1996). As a result, different nations have elevated the safeguarding of their cultural heritage from "state action" to "national policy." As per existing literature (Adane, Chekole, & Gedamu, 2019; Khan, Shafi, & Ahangar, 2018; Borissova, 2018; Ognjanović, Marinković, Šegan-Radonjić, & Masliković, 2019; He, Ma, & Zhang, 2017), preservation, promotion, and increased access opportunities are fundamentals of digitization. Also, these studies indicate that one of the advantages of the digitization of built heritage includes tourism and associated activities.

Digitization has its limitations too. Heritage and culture are intermediaries that link individuals to historical events or behaviors that generate genuine experiences. Authenticity is not an inherent characteristic of heritage customs; rather, it resides in the experiences that may be produced via the use of heritage artifacts (Manžuch, 2017). According to El-Said & Aziz (2022), virtual tours could never match the pleasure of physically experiencing a monument or site. Digitization encounters challenges in the form of insufficient funding, a dearth of suitable technology, and concern from authorities during the formulation of policy, plan, and process (Adane, Chekole, & Gedamu, 2019; Khan, Shafi, & Ahangar, 2018; Bountouri, 2017).

India, being one of the earliest civilizations, has a massive repository of historical legacy knowledge materials written in a variety of languages and scripts that have been handed down over the years. Conservation and preservation of this plethora of heritage monuments and sites were assigned to the Archaeological Survey of India, which was established during colonial rule. Later on, a few non-government organizations like the Indian National Trust for Arts and Cultural Heritage (INTACH) and the Indian Trust for Rural Heritage and Development (ITRHD) were voluntarily involved in the same business (Bakshi, 2016). Established by the Indian government in 1987, the Indira Gandhi National Centre for the Arts (IGNCA) uses various preservation techniques, tools, and technology for the digitalization of cultural heritage resources in a variety of formats, such as manuscripts, microfilms, microfiches, negatives, photographs, audio-visuals, etc. (Ahmad & Sharma, 2020). Converting built heritage into 3D models to preserve the built heritage and attract tourist footfall to boost tourism was initiated by Rajasthan (Singh, 2017) and later expanded to other states (Shekhar, 2017).

4.3 RESEARCH METHODOLOGY

This study was designed to understand and investigate the perceptions of Indian domestic tourists wanting to experience and learn about heritage monuments. The digitization of built heritage in India is still in its developing phase. Therefore, to gain a detailed understanding of the perceptions of Indian tourists on experiencing a monument digitally or physically is to be determined. This study employed a qualitative approach in which semi-structured interviews were conducted through open-ended questions. 21 respondents were selected through convenience sampling to achieve a theoretical saturation point. The base criteria for choosing these respondents were their exposure to digital heritage, i.e., these respondents had experienced digitized monuments (through AR and VR) at least once in their lifetime.

These respondents were traced at three well-known heritage sites in Delhi, i.e., Humayun's Tomb, Jantar Mantar, and Qutub Minar protected through the ASI (Archaeological Survey of India), a government agency, and were asked to describe their thoughts on the digitization of built heritage. The collected data was analyzed by employing MAXQDA software to analyze the responses and draw conclusions.

4.4 DISCUSSION

The relationship between heritage and tourism is complex. Heritage tourism destinations constantly face challenges related to heritage preservation, protection, safeguarding, management, and subsequent heritage promotion combined with viable tourism experiences. It is recognized that the core product of a heritage site is the visitor experience. Researchers from various disciplines have conducted extensive research on the heritage visitor experience, which has gained prominence as a research topic. Knowledge gains, education, participation, entertainment, active involvement, and engagement have been identified as the fundamental expectations that visitors/tourists have, along with the attractiveness of the heritage site and various other critical aspects (such as audio and visual communication, atmospherics, on-site engagement, information, etc.) that are found to influence the visitor heritage experience (Masberg & Silverman, 1996; De Rojas & Camarero, 2008).

Digitalization is trending worldwide, and its applications are explored in diverse fields of study. Governments across the world are exploring the various facets of digitization in the field of tourism. A similar wave of digitalization is being explored and emphasized by the Government of India (GoI) in businesses and the economy (Raj & Aithal, 2018). The Government of India (GoI) in particular has emphasized the digitization of the economy since 2014. The tourism industry in India has been at the forefront of digitization, has exponentially transformed, and has not only been involved in online hotel bookings and digital payments but has also prompted the development of new avenues in tourism. One such new product is virtual tourism, which has gained prevalence among Indian tourists in recent times (Kumar & Shekhar, 2020). As the concept is new, it is essential to understand the willingness of Indian tourists to experience tourism through the lens of digitization, especially in the case of heritage monuments, where preserving the monuments for the future is essential and sustainable consumption is a growing concern.

The onset of the COVID-19 pandemic triggered an opportunity for the proliferation of virtual travel, and the pandemic pushed individuals to explore this alternate form of travel and experience a place, which also required individuals to develop familiarity with technology (Talwar et al., 2022). But with the world getting back to normal in the post-COVID scenarios, it is necessary to investigate the tourist perception with regard to experiencing the heritage monuments. At the same time, understanding the pressing issue of sustainable consumption/experience of built heritage and preserving monuments for future generations is also a topic of interest that requires deliberation (Yeniası & Gökbulut, 2022). To achieve the objectives of our study, we have specifically asked the visitors to respond to four questions listed in the questionnaire: Q1: What is your preference in terms of experiencing the monument? Would you prefer a digitized experience/virtual tour, or do you prefer to visit physically the site of the monument? Kindly elaborate on the choice made. Q2: What are your thoughts on the effect of digitization of monuments as a method of preserving monuments digitally? Q3: In your opinion, what are the possible challenges that you feel you might face while experiencing monuments digitally? Q4: In your opinion, what are the possible advantages of digitizing monuments? Can digitization promote the sustainability of monuments and also promote tourism? Please elaborate.

4.4.1 EXPERIENCING THE HERITAGE MONUMENT: VISITOR PREFERENCES (IN-SITU EXPERIENCES OR DIGITIZED EXPERIENCES).

Built heritages are architectural marvels and provide a glimpse into the traces of past civilizations. These monuments are considered important for tourism activities, but at the same time, increased tourism activities can lead to the destruction of these sensitive archaeological sites of historical importance (Kurucu, 2021). Social and technological developments in recent times have changed the way we experience tourism. Digitalization is changing the traditional ways of experiencing tourism in recent times, and the digital preview of a destination's premises can provoke imagery before even physically experiencing the place (Lin et al., 2020; Lixing, 2020; Akhtar et al., 2021). Even though the literature is dominated by studies pointing in the direction that new technologies like VR have changed the way we experience tourism and are frequently used as a marketing tool (Yung & Khoo-Lattimore, 2019). These technological advancements and adaptations have seen a steep rise because of the COVID-19 pandemic, as visiting a destination physically was not possible (Chinazzi et al., 2020). But now that the world is returning to normal, the researchers believe a reevaluation of the perceived usefulness and relevance of digital technology must be conducted.

A word cloud generated (Figure 4.1) from a word frequency test on the responses to Q1 revealed that words such as "physical" and "visit" were stated more often, suggesting the respondents' inclination towards physically visiting a destination. Similarly, the higher frequency of words such as "see" and "experience" points towards the involvement of sensory memory.

FIGURE 4.1 Word cloud for Q1.

Source: Primary data for Q1.

Given the complexity of consumer behavior, the researchers are intrigued to understand the perspectives of the tourists, whether they intend to experience a historical monument in situ or they prefer experiencing a monument in Virtual Reality (VR) in the post-pandemic world. Therefore, to understand the viewpoints of visitors visiting a heritage monument, the first probing question was to understand if the visitors were willing to engage in virtual tours or if visiting a monument physically was a preference for them to experience the historical monument.

Interestingly, the majority of them responded that they would like to visit the destination physically. A few of the collected responses, given verbatim, are provided below:

"I would prefer to visit in person. There is something intangible that we get to experience at a destination."

"I would like to visit the site of the monument physically. Because you can feel the monument and the time it was built by walking in and around it. Virtually, you can only see it; you cannot feel it."

"Physically at the monument's location. It involves the physical attachment to the monument. The beauty of the monument as perceived by my senses aids in my understanding and development of my perception. A digital experience is already made through someone else's perception."

"I prefer visiting the site physically because it allows me to examine the monument details more closely and vividly."

"I would like to experience the monuments physically as it gives me a chance to embrace the beauty of the monuments, whereas a digitized experience would only inform an individual and not create memories, which one wants when we visit a place of interest."

"Physically on site is better because you get a lot of time to explore the place, and you can feel the exact environment of the place too, which is not possible in virtual mode."

On examining the responses collected, it can be appropriately inferred that most of the visitors are willing to experience and visit the historical monuments physically. Various studies on heritage tourism indicate that visitor experiences are usually shaped by the affective and cognitive experiences (Prentice et al., 1998) gained at the historic site, which majorly relies on the quality of interpretive media made available on these sites and could culminate in "sense of place" formation (Stewart et al., 1998). Since it is widely accepted that visiting a heritage site could be classified as an informal learning activity, if combined with heritage-site visiting activities based on the sense of place theory, it could develop the visitor's interest and also result in visitors learning about these sites (Chang et al., 2015).

4.4.2 DIGITIZATION AS A METHOD OF PRESERVING MONUMENTS

The most noteworthy category of the cultural heritage of a destination is its historic or architectural monuments, and the preservation and conservation of these architectural marvels have become a universal priority (Pavlidis, 2007). Urbanization and neglect of historic sites and buildings have been the core reasons that have led to vernacular architecture, culture, and societal values losing their sheen (Fadli, 2013). Preservation of historic sites in urban settings is necessary and relates to the concept of sustainability, as these sites are indispensable reflections of the past. The process of built heritage conservation is a continuous process, and one method of achieving it is the "digitization of heritage," which involves the application of the physical 3D survey technique for cataloging and archiving the architectural components of built heritage (Fadli & AlSaeed, 2019).

The word frequency graph generated from the responses to Q2 (Figure 4.2) highlights the fact that the respondents believed that digitization has great potential to preserve physical monuments. This can be proved by the higher frequencies of words like "digitally" and "digitization," and positive words like "good" and "great."

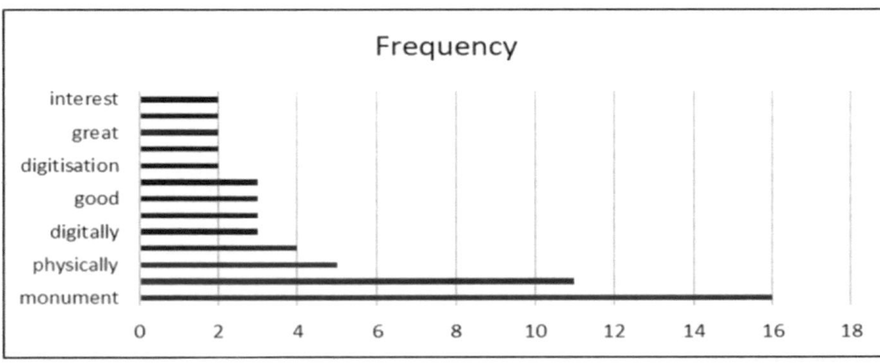

FIGURE 4.2 Word frequency graph for Q2.

Source: Primary data for Q2.

As the majority of participants have limited exposure to digital technologies and thus show little enthusiasm towards experiencing a monument virtually, most of them showed enthusiasm when visiting the heritage site. However, when participants were asked to ponder the idea of digitizing a heritage site as a means of preserving it, most of them were enthusiastic about the idea of digitizing the monument and preserving it for future generations, but they also expressed their opinion on safeguarding, protecting, and preserving the physical structures as being of utmost importance. A few responses collected from the respondents supporting the notion of "digitization as a method of preserving monuments" are mentioned below:

"It will preserve the monuments in their true form, as we know monuments decay with time. So even after a century, we could have access to how the original sites looked."

"Digitization of monuments is a new method of their preservation, but in order to maintain the structure physically, one needs to visit and look up renovations needed with the passage of time."

"I worry that preservation through digitization of monuments might lessen or dilute the efforts of preserving them physically, which can lead them into ruins."

"It is indeed a great step towards preservation as it can aid travel fantasies and spur interest in unknown locations. It can help preserve fragile monuments by decreasing the footfall and allowing people to experience such monuments remotely."

"It sounds interesting, and I feel like it's a nice way to portray some monuments that are either destroyed over time or don't exist anymore. It can give us a glimpse of what used to exist and its glory at that time."

"Yes, I agree with this method of preserving monuments digitally as sometimes, with time passing, the original grandeur may be lost gradually."

4.4.4 PERCEIVED CHALLENGES WHILE EXPERIENCING MONUMENTS DIGITALLY

Bringing people back to the site and initiating a dialogue is the ultimate goal of cultural heritage and cultural sites (Li, 2010). However, the growing influence of technology has influenced processes, objects, and services to transform themselves. COVID-19 accelerated the growing influence of technologies within societies and influenced the need to solve human problems through technology, with cultural tourism being no exception.

The word cloud generated from the responses to Q3 points towards the finding that digitization of monuments may be challenged by "technical issues" and a "lack" of resources like network infrastructure or bandwidth, which may result in destinations losing their authenticity and physical feel (Figure 4.3).

Over the past few decades, different technologies have been used more and more as a way to show and explain culture (Othman et al., 2011). However, these digital technologies didn't become widely used in the West until the 1990s (He et al., 2017). Digitization of architectural heritage depends on factors like size, budget, and applicability, and it is important to consider these factors before choosing an appropriate digitization method (Pavlidis, 2007). Technology readiness and acceptance are also vital factors that influence tourist behavior. It can also be understood from the responses

FIGURE 4.3 Word cloud for Q3.

Source: Primary data for Q3.

collected that connectivity issues and internet facilities influence visitor behavior at a built heritage site. A study by Amaro and Duarte (2015) also supports the idea that digitalization in tourism depends on user compatibility, convenience, and perceived risk.

A few of the responses collected from the respondents are given below:

"The sense of being present somewhere that you get when you are there would be lost. It cannot make you feel alive while experiencing it digitally. It's not just the monument that you experience when you visit the place. You also meet people and get to experience new things, such as different types of street food."

"As far as the limitations are concerned, they may be seen at both levels: at the level of digitizing the monuments as well as at the level of the touristic experience. The lack of proper funding, lack of trained personnel, etc., may be seen at the level of digitizing the monuments. On the other hand, at the level of experiencing a monument digitally, the access to such monuments will be limited to the literate class only, and making the masses aware of them will become a challenging task."

"Sometimes we can face unwanted network issues or we can also face struggles in promoting it in remote areas due to a lack of connectivity and facilities."

The visitors are of the opinion that digitization can be considered a challenge in a developing country like India, where not everyone has access to smartphones and a majority of people are not aware of or trained to use these technologies. Research from the past has also shown that awareness and education about tourism and technology affect how people feel about virtual reality and digitization, especially in both developing and developed countries (Akhtar et al., 2021). Although digitization and the application of digital technologies can open new avenues for experiencing cultural heritage, at the same time, they could also interfere with visitor experiences if applied without careful thought and understanding of visitors' acceptance and expectations (Liu, 2020).

4.4.5 DIGITIZATION AS A MODE OF PROMOTING SUSTAINABILITY OF MONUMENTS AND PROMOTING TOURISM

The durability of the structure of historic buildings is a major factor that contributes to their survival (Fadli & Sibley, 2008). The well-being of these buildings is often compromised to make way for the conveniences and practicalities of modern life.

Sustainable heritage usually addresses two issues: environmental protection and community participation, which can be counted as a social dimension. Several studies recommend the support and engagement of local communities to conserve built heritage and support tourism development (Yuan et al., 2019). Our respondents have also supported the role and engagement of local communities, and the sustainable livelihoods these sites generate by creating employment are also part of sustainability (Lwoga, 2016). As pointed out by one of the respondents, digitization and virtual tours could have negative impacts on the livelihoods of the local community.

The word cloud generated from the retrieved data for Q4 revealed that "digitization" is thought to be one of the best methods to increase the "sustainability" of the "monuments." Higher frequencies for positive words such as "yes," "employment," "accessibility," "advantage," and "preserve" point towards the benefit of going "digital" (Figure 4.4).

Socially responsible practices are essential for heritage and environmental protection. To ward off the negative effects of tourism, environmentally responsible behavior among tourists is important (Chow et al., 2019). Limiting visits to heritage sites, making donations being willing to pay for preservation, and volunteering to support the preservation of a heritage site are all examples of pro-heritage behaviors (Buonincontri & Ramkissoon, 2017). From the responses collected, it is evident that sensitizing visitors

FIGURE 4.4 Word cloud for Q4.

Source: Primary data for Q4.

to experience the heritage site could help in preserving the monument for future generations. Another participant in the study finds digitization to be a method of crowd management at historic monuments.

"I don't think it can promote tourism because it will not benefit economically since the area around the sites generates a lot of jobs, and the digitization of the sites will negatively affect those people in terms of employment, growth, and sustainability.

Digitizing the monuments does have some advantages, and yes, sustainable preservation is one of them. It also has advantages in expanding the field of knowledge and can make those places accessible to history enthusiasts, which has remained unpopular until now.

The advantage is totally in terms of attracting tourists and encouraging tourism. Another advantage of digitization is accessibility. It can also help in creating a whole picture of the monument and its subsequent development and changes, which can promote or sensitize people about preserving monuments and giving them value.

However, sustainability is a larger question within heritage tourism that needs to be addressed while experiencing a monument physically as well as digitally.

The advantages of digitization for monuments can vary widely. One can sit at home and learn about monuments of their choice without being physically present. Also, it is helpful from a school student's educational point of view if teachers start with demonstrations on projectors showing different monuments worldwide. Yes, on a mass level, it promotes tourism as one can plan a trip and pick up locations before visiting them, or on maps, they can be of great help for travelers as well.

The advantage of digitizing monuments is that it is going to facilitate crowd management in addition to the preservation of monuments. It is also an economical option."

In terms of teaching and learning, the digitization of monuments can stimulate students' understanding and explain histories that are difficult to infer or understand by using traditional instructional methods (Noh et al., 2009).

4.5 CONCLUSION

India has huge untapped potential in terms of cultural heritage resources. There is tremendous scope for creating immersive experiences through optimal usage of technology and digital interfaces. The heritage management programs and governance models should explore the possibilities of harnessing the new age technologies to create sustainable models of preserving and promoting the cultural heritage of India, especially the built heritage and monuments. The results of the qualitative survey confirm that heritage consumption is laden with experiential value, and the tourists are of the opinion that the experiences they gain or have gained in the past while visiting the heritage monuments have an emotional and sensorial ring to them. Preserving, conserving, and transferring cultural heritage for future generations are among the biggest challenges we face as contemporary societies. In this context, various agencies, including governments and non-profit agencies, are playing an important role in conserving and promoting historical sites as places to learn from our past.

This chapter aims to understand the acceptance of the "digitization of monuments" and its subsequent transition into viable tourist experiences among the urban Indian populace. The authors found that despite the upsurge of technologies and their application in various fields of study, digitization of monuments and visitors wanting to experience the monument virtually have not gained prominence in India. Our studies are in line with the study conducted by El-Said & Aziz (2022), which included respondents from over 40 countries and substantiated the fact that virtual tours could never replace the experience of visiting a site. Even though experiences matter the most in terms of a touristic activity taken up by a tourist, digitization could provide ancillary support to enhance the visitor experience and also help preserve monuments.

However, digitization remains one of those 21st-century technologies that could not only help in preserving the digitized version of the built heritage but could also assist the intangible elements of cultural heritage like the cultural and social practices of its inhabitants. These findings also support the findings of Fadli & AlSaeed (2019), which elaborate that digitalization is finding its relevance as a method of preventive conservation by implementing robust tools to link the physical and digital worlds for proper documentation and management of cultural heritage (Sánchez-Aparicio, 2020). The work can be employed as a future reference by organizations working towards the digitization and preservation of India's built heritage and could galvanize sustainable tourism practices at these heritage sites.

KEYWORDS

- **application programming interface**
- **Archaeological Survey of India**
- **augmented reality**
- **built heritage digitization**
- **conservation**
- **COVID-19**
- **sustainability**
- **virtual reality**

REFERENCES

Adane, A., Chekole, A., & Gedamu, G. (2019). Cultural Heritage Digitization: Challenges and Opportunities. *International Journal of Computer Applications, 178*(33), 1–5. https://doi.org/10.5120/ijca2019919180.

Ahmad, A., & Sharma, S. (2020). Sustainable Digital Preservation and access of Heritage Knowledge in India. *DESIDOC Journal of Library & Information Technology, 40*(05), 321–325. https://doi.org/10.14429/djlit.40.05.15822.

Akhtar, N., Khan, N., Mahroof Khan, M., Ashraf, S., Hashmi, M. S., Khan, M. M., & Hishan, S. S. (2021). Post-COVID 19 tourisms: Will digital tourism replace mass tourism? *Sustainability, 13*(10), 5352. https://doi.org/10.3390/su13105352.

Amaro, S., & Duarte, P. (2015). An integrative model of consumers' intentions to purchase travel online. *Tourism Management, 46*, 64–79.

Bachi, V., Fresa, A., Pierotti, C., & Prandoni, C. (2014). The digitization age: Mass culture is quality culture. Challenges for cultural heritage and society. *Digital Heritage. Progress in Cultural Heritage: Documentation, Preservation, and Protection*, 786–801. https://doi.org/10.1007/978-3-319-13695-0_81. Lecture Notes in Computer Science, vol 8740. Springer, Cham.

Bakshi, S. I. (2016). Digitization and digital preservation of cultural heritage in India with special reference to IGNCA, New Delhi. *Asian Journal of Information Science and Technology, 6*(2), 1–7.

Balandin, S. (2013). "Preface of the 13th FRUCT conference proceeding," 2013 13th Conference of Open Innovations Association (FRUCT), Petrozavodsk, Russia, 2013, pp. VII-IX, doi: 10.23919/FRUCT.2013.8124242.

Borissova, V. (2018). Cultural Heritage digitization and related intellectual property issues. *Journal of Cultural Heritage, 34*, 145–150. https://doi.org/10.1016/j.culher.2018.04.023.

Bountouri, L. (2017). 3 - Digitization. In L. Bountouri (Ed.), *Archives in the Digital Age* (pp. 29-36). Chandos Publishing. https://doi.org/https://doi.org/10.1016/B978-1-84334-777-4.00003-7.

Buonincontri, P., Marasco, A., & Ramkissoon, H. (2017). Visitors' experience, place attachment and sustainable behavior at cultural heritage sites: A conceptual framework. *Sustainability, 9*(7), 1112.

Chang, Y. L., Hou, H. T., Pan, C. Y., Sung, Y. T., & Chang, K. E. (2015). Apply an augmented reality in a mobile guidance to increase sense of place for heritage places. *Journal of Educational Technology & Society, 18*(2), 166–178.

Chinazzi, M., Davis, J. T., Ajelli, M., Gioannini, C., Litvinova, M., Merler, S., & Vespignani, A. (2020). The effect of travel restrictions on the spread of the 2019 novel coronavirus (COVID-19) outbreak. *Science (New York, N.Y.), 368*(6489), 395–400.

Chow, A. S., Ma, A. T., Wong, G. K., Lam, T. W., & Cheung, L. T. (2019). The impacts of place attachment on environmentally responsible behavioral intention and satisfaction of Chinese nature-based tourists. *Sustainability, 11*(20), 5585.

Chowdhury, G. (2002). Tourism and Digitization of Cultural Heritage. In *Proceedings of the 5th International Conference on Asian Digital Libraries: Digital Libraries: People, Knowledge, and Technology (Vol. 58, No. 3, pp. 258–283).*

De Rojas, C., & Camarero, C. (2008). Visitors' experience, mood and satisfaction in a heritage context: Evidence from an interpretation center. *Tourism Management, 29*(3), 525–537.

El-Said, O., & Aziz, H. (2022). Virtual tours a means to an end: An analysis of virtual tours' role in tourism recovery post COVID-19. *Journal of Travel Research, 61*(3), 528–548. https://doi.org/10.1177/0047287521997567.

Fadli, F. (2013, November). Digitizing vanishing heritage in the MENA region: Case of Qatar. In *Qatar Foundation Annual Research Forum (Vol. 2013, No. 1, pp. SSHP-029)*. Qatar: Bloomsbury Qatar Foundation Journals.

Fadli, F., & AlSaeed, M. (2019). Digitizing vanishing architectural heritage; The design and development of Qatar historic buildings information modeling [Q-HBIM] platform. *Sustainability, 11*(9), 2501.

Fadli, F., & Sibley, M. (2008). The Historic hammāms of Cairo: Safeguarding a vanishing heritage. *Journal of Architectural Conservation, 14*(3), 59–80.

Fiorillo, F., Remondino, F., Barba, S., Santoriello, A., De Vita, C. B., & Casellato, A. (2013). 3D digitization and mapping of heritage monuments and comparison with historical drawings. *ISPRS Annals of the Photogrammetry, Remote Sensing and Spatial Information Sciences, II-5/W1*, 133–138. https://doi.org/10.5194/isprsannals-ii-5-w1-133-2013.

Gantait, A., Mohanty, P., & Swamy, G. A. (2018). Conservation and Management of Indian Built-Heritages: Exploring the Issues and Challenges. *South Asian Journal of Tourism and Heritage, 11*(1), 5–21.

He, Y., Ma, Y. H., & Zhang, X. R. (2017). "Digital heritage" Theory and innovative practice. *International Archives of the Photogrammetry, Remote Sensing & Spatial Information Sciences*, XLII-2/W5, 335–342.

ICOMOS International Committee on Cultural Tourism. (2002). ICOMOS international cultural tourism charter: Principles and guidelines for managing tourism at places of cultural and heritage significance. *International Council on Monuments and Sites, ICOMOS International Cultural Tourism Committee.*

Khan, N. A., Shafi, S. M., & Ahangar, H. (2018). Digitization of cultural heritage. *Journal of Cases on Information Technology, 20*(4), 1–16. https://doi.org/10.4018/jcit.2018100101.

Kumar, S., & Shekhar. (2020). Digitalization: A strategic approach for development of tourism industry in India. *Paradigm, 24*(1), 93–108.

Kurucu, M. (2021). An Investigation of the Use of Virtual Reality in Archaeological Heritage Areas in the Context of Conservation and Tourism. *Researches on Urban and Architectural Design Studies*, 152.

Li, W. (2010). Museum culture and new media communication. In L. L. Jia (Ed.), *Preservation and Presentation of Cultural Heritage in Digital Age* (pp. 92–109). Beijing: Culture and Art Publishing House.

Lin, L. P. L., Huang, S. C. L., & Ho, Y. C. (2020). Could virtual reality effectively market slow travel in a heritage destination? *Tourism Management, 78*, 104027.

Liu, Y. (2020). Evaluating visitor experience of digital interpretation and presentation technologies at cultural heritage sites: A case study of the old town, Zuoying. *Built Heritage, 4*(1), 14.

Lixing, Z. (2020). Impact of the COVID-19 Outbreak on the World and the Inspirations. *Global Journal of Management and Business Research, 20*(B5), 29–37.

Lwoga, N. B. (2016). Tourism employment and local residents' engagement in the conservation of the built heritage in Zanzibar Stone Town in Tanzania. *WIT Transactions on Ecology and the Environment, 201*, 43–55.

Mannoni, B. (1996). Bringing museums online. *Communications of the ACM, 39*(6), 100–105. https://doi.org/10.1145/228503.228524.

Manžuch, Z. (2017). Ethical issues in digitization of cultural heritage. *Journal of Contemporary Archival Studies, 4*(2), 4.

Masberg, B. A., & Silverman, L. H. (1996). Visitor experiences at heritage sites: A phenomenological approach. *Journal of Travel Research, 34*(4), 20–25.

Mohanty, P., Hassan, A., & Ekiz, E. (2020). Augmented reality for relaunching tourism post-COVID-19: Socially distant, virtually connected. *Worldwide Hospitality and Tourism Themes, 12*(6), 753–760. https://doi.org/10.1108/WHATT-07-2020-0073.

Noh, Z., Sunar, M. S., & Pan, Z. (2009). A review on augmented reality for virtual heritage system. In *Learning by Playing. Game-based Education System Design and Development:*

4th International Conference on E-Learning and Games, Edutainment 2009, Banff, Canada, August 9–11, 2009. Proceedings 4 (pp. 50–61). Springer Berlin Heidelberg.

Ognjanović, Z., Marinković, B., Šegan-Radonjić, M., & Masliković, D. (2019). Cultural Heritage Digitization in Serbia: Standards, policies, and case studies. *Sustainability, 11*(14), 3788. https://doi.org/10.3390/su11143788.

Othman, M. K., Petrie, H., & Power, C. (2011). Engaging visitors in museums with technology: Scales for the measurement of visitor and multimedia guide experience. In *Human-Computer Interaction–INTERACT 2011: 13th IFIP TC 13 International Conference, Lisbon, Portugal, September 5–9, 2011, Proceedings, Part IV* (pp. 92–99). Springer Berlin Heidelberg.

Packer, J., & Ballantyne, R. (2016). Conceptualizing the visitor experience: A review of literature and development of a multifaceted model. *Visitor Studies, 19*(2), 128–143.

Pavlidis, G., Tsiafakis, D., Koutsoudis, A., Arnaoutoglou, F., Tsioukas, V., & Chamzas, C. (2007). Preservation of architectural heritage through 3D digitization. *International Journal of Architectural Computing, 5*(2), 221–237. https://doi.org/10.1260/1478-0771.5.2.222.

Prentice, R., Guerin, S., & McGugan, S. (1998). Visitor learning at a heritage attraction: A case study of discovery as a media product. *Tourism Management, 19*(1), 5–23.

Raj, K., & Aithal, P. S. (2018). Digitization of India-impact on the BOP sector. *International Journal of Management, Technology, and Social Sciences (IJMTS), 3*(1), 59–74.

Ratrrout, A., Zidan, N., Abulhouf, A., & Sabha, M. (2006). Palestine Heritage Digitization: Collaborative distributed system approach. *Revue d'Interaction Homme-Machine, 7*(1).

Remondino, F., Barazzetti, L., Nex, F., Scaioni, M., & Sarazzi, D. (2011). UAV photogrammetry for mapping and 3D modeling – Current status and future perspectives. *The International Archives of the Photogrammetry, Remote Sensing and Spatial Information Sciences, XXXVIII-1/C22*, 25–31. https://doi.org/10.5194/isprsarchives-xxxviii-1-c22-25-2011.

Sánchez-Aparicio, L. J., Masciotta, M. G., García-Alvarez, J., Ramos, L. F., Oliveira, D. V., Martín-Jiménez, J. A., & Monteiro, P. (2020). Web-GIS approach to preventive conservation of heritage buildings. *Automation in Construction, 118*, 103304.

Shekhar, D. (2017). Here's how technology is helping conservationists to document, restore and maintain our history. Retrieved from: https://economictimes.indiatimes.com/magazines/panache/heres-how-technology-is-helping-conservationists-to-document-restore-and-maintain-our-history/articleshow/58180268.cms?from=mdr (accessed on 18 June 2024).

Singh, R. (2017). Seven Rajasthan Monuments Get Digital Protection: Jaipur News – Times of India. Retrieved from: https://timesofindia.indiatimes.com/city/jaipur/seven-raj-monuments-get-digital-protection/articleshow/57741808.cms (accessed on 18 June 2024).

Stewart, E. J., Hayward, B. M., Devlin, P. J., & Kirby, V. G. (1998). The "place" of interpretation: A new approach to the evaluation of interpretation. *Tourism Management, 19*(3), 257–266.

Talwar, S., Kaur, P., Nunkoo, R., & Dhir, A. (2022). Digitalization and sustainability: Virtual reality tourism in a post-pandemic world. *Journal of Sustainable Tourism, 31*(11), 1–28.

UNESCO. (2003). Charter on the Preservation of Digital Heritage. Retrieved from: https://unesdoc.unesco.org/ark:/48223/pf0000179529 (accessed on 18 June 2024).

UNESCO. (n.d.). What is Intangible Cultural Heritage? Retrieved from: https://ich.unesco.org/en/what-is-intangible-heritage-00003

Yeniasır, M., & Gökbulut, B. (2022). Effectiveness of usage of digital heritage in the sustainability of cultural tourism on islands: The case of Northern Cyprus. *Sustainability, 14*(6), 3621.

Yuan, Q., Song, H., Chen, N., & Shang, W. (2019). Roles of tourism involvement and place attachment in determining residents' attitudes toward industrial heritage tourism in a resource-exhausted city in China. *Sustainability, 11*(19), 5151.

Yung, R., & Khoo-Lattimore, C. (2019). New realities: A systematic literature review on virtual reality and augmented reality in tourism research. *Current Issues in Tourism, 22*(17), 2056–2081.

CHAPTER 5

THE EVOLUTION OF THE IMAGE OF ANDALUSIA IN TOURISM: FROM THE FOREIGN VISION TO THE INTEGRATION OF THE ROMANTIC CLICHÉ IN CONTEMPORARY PROMOTIONAL ACTIVITY

JOSÉ DAVID ALBARRÁN PERIÁÑEZ

Independent Researcher and Consultant, PhD in Geography, Complutense University of Madrid, Spain

ABSTRACT

Throughout the 20th century, the world tourism market witnessed a transformation by becoming increasingly competitive. Consequently, and because the territories attempted to stand out in a context characterized by the survival of 19th-century aesthetics, the iconographic bases of Romanticism gradually modified and complemented themselves until they conformed with the array of tourist images in the globalized reality of the contemporary world (Barthes, 1957; Lerivray, 1975; Torres, 2017; Méndez, 2017). This was Andalusia's case since the origins of its promotional image as interstate travel consolidated itself during the 17th and 18th centuries. At that time, the European upper classes were attracted by the decadence and exoticism of this depressed land.

In this study, the development of a semiotic analysis of the region's image as it was projected by enlightened travelers through tourist promotion today

Dynamics of Heritage Management and Tourism Development: Trends, Practices, and Approaches.
Surabhi Srivastava & Ruchika Kulshrestha (Eds.)

aims to confirm that the picturesque has become an object of commercialization by its protagonists. In this sense, the establishment of the picturesque as the dominant image in tourism policy as implemented by public authorities to date contributes to various reflections on a subject for which scientific research remains scarce. Through a methodology of semiotic analysis, the information included in multiple visual supports is systemized and interpreted. These techniques apply to literary and pictorial references carried out by travelers, engravers, and cartoonists of the 18th and 19th centuries. The results of this work prove the unfolding evolution of perception from external conceptualizations to the assimilation of 19th-century patterns by the authorities in charge of their territory.

5.1 INTRODUCTION

In a context like the current one, which is strongly dominated by the visual, the tourism industry has established itself as one of the most important sectors of the global economy. Thanks to its ability to generate foreign currency and business activity, tourist destinations compete "on the basis of their perceived image in relation to competing destinations" (San Martín et al., 2006, p.70). There is a broad consensus on the idea that the tourist image of a destination is a critical factor in the process involved in its eventual selection (Goodrich, 1978; Woodside & Lysonski, 1989; Andreu et al., 2000; Perelló, 2006; Edelheim, 2007; San Martín & Rodríguez, 2008; Picón et al., 2013; Castro & Rodrigues, 2017). The organizations responsible for tourism promotion do their very best, therefore, to produce an identity capable of achieving a prominent position in the perception of its consumers and succeeding in the market (Santos, 2008). They do this through positive images that incite travel while differentiating themselves from other places (Novo et al., 2013). This has incited the proliferation of destination brands, the creation of positioning strategies, and the large-scale production of seductive and evocative images by the same agents responsible for local tourism management.

In this way, the actions implemented by destinations to attract tourists have intensified in recent decades, for which a selection of elements has proven worthy in allowing the destination to be recognized by consumers (Rodríguez et al., 2012; Novo et al., 2013). The formation of a tourist image, therefore, is based on a small number of impressions selected from a much larger set. This simplification that generates the creation of the tourist destination image may lead to the generation of certain stereotypes. These are largely due to the influence generated by these images during the Romantic period, a moment marked by the initiation of interstate travel in Europe and the great world expeditions (Barthes, 1957; Lerivray, 1975; Torres, 2017; Méndez, 2017). The survival of this 19th-century aesthetic through to the present day has favored tourist representations founded on the iconographic bases of Romanticism. Modern tourists travel in search of utopia in the authenticity of past times (Dachary & Arnáiz, 2009). Consequently, based on the categorization of tourist images proposed by Gunn (1988), various authors have argued that the current induced image, that is, the one created by public organizations responsible for destination-based tourist promotion, is based on a universal image of itself – the one that is unconsciously fed into the social imagination and legitimized over a long period of time.

The perceptions individuals have about different territories, shaped by culturally defined contexts, are therefore influenced or reinforced by the agents involved in shaping the promotional image of these destinations. And they, through the induced image, arouse interest in traveling and ultimately determine which sights will become essential to visit, observe, or photograph (Urry, 1990).

Nevertheless, the induced tourist images are far from the true reality of the territories they represent. On the one hand, the myth of paradise and the ambition of 19th-century writers to generate impact and emotion in their readers (Carrère-Lara, 2001) favored the publication of numerous stories throughout the 19th century, that were based on clichés and generalized prejudices about the territory they described. On the other hand, the search for prominent positions by the agents in charge of the tourism market contributes to the construction of promotional patterns trying to satisfy the external demand through images aligned with the shared idea of the territories and the broadcasts of tourism from their sites. Therefore, an increase in the number of public promotions during the last decades has translated into a gradual modification of the traditional 19th-century imagery to make these sites more profitable for tourism (Edelheim, 2007; Martín, 2003), adapting themselves to more contemporary tourist demands (Santillan, 2010).

In this continuous aspiration to generate images capable of guaranteeing destination recognition by consumers (Rodríguez et al., 2012; Novo et al., 2013), heritage assets have been imposed as differentiating elements, since they respond to well-established perceptions in the collective imagination and allow the generation of a unique image of each territory (Castro & Rodrigues, 2017). In this way, the assets that gave cohesion to the imagination of the past have been rescued and converted into products for tourist consumption. This is not only from their symbolic projection through the creation of a strengthened image inside the context of tourism marketing but also through physical adaptation with the conditioning functionality of these same images. In this way, heritage has become an attractive distinction for tourist activity (Troncoso, 2013), in a process accentuated in those territories with a distinctive patrimonial heritage and that have undergone some type of colonization process throughout their history (Dachary & Arnaiz, 2009; Albarrán, 2021).

In this way, the present work evaluates the way in which the image of the region of Andalusia, Spain, has evolved from when it was projected by the enlightened traveler until the tourist promotion carried out by public officials at the current time, detecting patterns and identifying common elements. The development of tourism in this region began with interstate travel in the 17th century when the European upper classes began regular visits to the area, attracted by the decadence and the exoticism of a depressed territory with African traits. Since then, the image of Andalusia has come to stand as the most representative and attractive in a country that was formerly considered brutal and cruel. By the 18th century, it had become one of the essential destinations for the elites of the continent. By this time, Andalusian had become synonymous with everything Spanish (García, 1995; Carrère-Lara, 2001; Garayo, 2001; Calatrava, 2018) through an iconography that would later be reinforced by large-scale tourist promotion (McDonald, 2015). Thus, after the first section focused on the theoretical study of the subject, the methodology applied to the research is presented. Subsequently, an analysis of the region's universal and induced images is developed. And finally, a section is dedicated to the development of a discussion on the results and the conclusions.

5.2 OBJECTIVE AND METHODS

The main objective of the current work is to move forward in determining the degree to which Victorian discourses became established as the dominant image in Andalusian tourism policy from the 17th to the 20th centuries. In so doing, the intent is to demonstrate the importance of generating more effective and functional endogenous discourses, distanced from the picturesque, which on many occasions have become the object of commercialization by the local administrations themselves.

To fulfill this objective, a study of the universal image of the region will first be developed through bibliographic review techniques analyzing literary, pictorial, or graphic profiles that helped build the common component of the image of Andalusia as portrayed in the collective imagination between the 17th and 19th centuries. This study allows the determination of which principal elements define the universal image of the territory based on the documents generated in the majority of cases by external agents and which are understood as primary sources.

On the other hand, the promotional materials edited by the public administrations in charge of the subject matter since the beginning of the 20th century are analyzed based on a methodology of semiotic analysis that is iconographic, allowing the identification and systemization of visual information included in tourism posters. This study is carried out by firstly considering the work carried out by state-level institutions and, consequent to the consecration of the Autonomic States, the publications published by the Junta de Andalucía. This analysis focuses on generic marketing and promotional campaigns, and not on material or publications about specific segments or destinations.

The image content analysis methodology allows the encoding of the elements being identified. It represents a process that, according to Picazo & Moreno-Gil (2017), is essential in evaluating the image being projected by tourist destinations, having been used repeatedly at the academic level (Buck, 1977; Thurot & Thurot, 1983; Albers & James, 1988; Dann, 1996; Dilley, 1986; Pritchard & Morgan, 2001; Urry, 2002; Edelheim, 2007; Singh & Lee, 2009; Milman, 2011; Donaire & Galí, 2011; Matteucci & Önder, 2018). Similarly, a subsequent semiological study allows us to analyze the meaning that underlies official tourist advertisements (Hernández, 2008), through their content, their composition, and their combination of both content and composition in order to communicate messages about the places they describe (Jenkins, cited by Edelheim, 2007).

A subsequent comparison proves how the evolution unfolded from the external perception to the assimilation of 19th-century patterns by the authorities in charge of their territory.

5.3 RESULTS

5.3.1 ANDALUSIA DURING THE ENLIGHTENMENT PERIOD

The consolidation of interstate travel in the 17th century entailed the transformation of the European perception of Spain. This country, previously resembling a violent and cruel territory, would later secure itself as one of the most indispensable destinations for the enlightened elites of the continent. Despite the fact that the philosophical or political contributions of other countries may have been considered of greater significance (Guerrero, 1990), Spanish decadence generated an occurrence of great interest, and the causes of its decline became a commonly broached subject in many travel stories (Guerrero, 1990; Fuertes-Majon, 2004). The state of its roads and infrastructure, as well as the decrepit situation of its inns, contributed to creating a decadent image for foreigners at the dawn of their inclusion in traveler circuits. Also, despite the absence of a defined itinerary for the journey through Spanish lands, by the end of the 18th century, Andalusia had become a region monopolizing such great prominence (Guerrero, 1990), that many travelers reduced their visits to Spain to a tour of only this territory (Bernal, 1985).

Although the image projected onto Andalusia during the Enlightenment is complex and rich in nuances, it encapsulated certain approaches that were common to most texts and engravings. These referred to desert and wild territory, where the forms of popular architecture were attributed to its Islamic heritage. Also significant are the allusions to ancient ruins and the townships that had colonized the region (Bourgoing, 1808; Peyron, cited by García, 1999; Bourgoing, cited by Soler, 2013). The critical connection to agricultural labor by the inhabitants of these lands throughout the entire Enlightenment resulted in an almost generalized opinion about the lazy character and personality typical of the Andalusian population, in a collective characterization responding to, with only some exceptions, that of an ignorant and violent society. These qualifiers were attributed by travelers to the heritage that the eastern peoples had left to the Andalusians. This belief, together with the danger of an encounter with thieves and bandits, compounded the adventure of the trip for later romantic travelers (Bernal, 1985).

5.3.2 THE SHAPING OF THE ROMANTIC IMAGE OF ANDALUSIA

At the beginning of the 19th century, Andalusia was still the region that exhibited the most genuine representation of Spain. This was the gateway to Africa (Galant, 2018), which meant that it was perceived as an exotic territory, combining the most typical Arab traditions with the medieval Christian world. These same aspects that during the Enlightenment had made visiting this region unattractive now consolidated it as an ideal destination for the upper classes that had rediscovered popular culture as an object of interest to explore.

Its themes were handled in a more symbolic and aesthetic way, inside which the wild, the Middle Ages, or the culture of the Far East were constantly exalted. This caused the arabesque archetype to be present in a large part of the written accounts, as well as in the paintings and engravings that were produced during the 19th century. This, in a context inside which representations of the Eastern world were becoming more common in Western art and literature, while the treatment of the Muslim evolved most prominently inside the projected image of Andalusia. This interest would plant the seed of the romantic image of Andalusia, presided over by "the three great sanctuaries of Islamic art: the Alhambra in Granada, the Mosque of Córdoba, and the Alcázares of Seville" (López, 2001:22), where orientalism impregnated all. In addition, rugged landscapes started becoming recognized as a refuge for passionate characters in an evolution of the enlightened idea of primitivism and brutality (Fernández et al., 2010; Bernal, 1985; López, 1988) (Figure 5.1).

FIGURE 5.1 Graphic representations of Andalusia during the 19th century (Chapel in the Great Mosque, Cordova, David Roberts, lithography, 1837; The Moorish tower at Seville, called the Giralda, David Roberts, oil on board, 1833; Patio de losArrayanes, Gustave Doré, xylography, 1864).

Source: Media Art, www.medinaarts.com. Sotheby's, sothebys.com. Descubrir el Arte, www.descubrirelarte.es.

However, many of the projections that were carried out on Andalusia during this century focused on the merely picturesque, thereby making references to the typical elements of North and Central African exoticism – such as deserts or camel bandits – a commonality. And this is why most of these authors stayed in the region for only short periods of time; they did not know the language, they hardly related to the locals other than as servants of their lodgings, and their visits centered on bullrings, monuments, and festivals which had been specially prepared for them (Bernal, 1985). In addition, travelers already arrived laden with prejudices that conditioned their attitudes during their trips, which had on many occasions been taken from previous literature, particularly from the Spanish Golden Age, which was in vogue during the Romantic period.

It was not until the end of the century when the progress of the railway and the creation of the Civil Guard to end insecurity on the roads began favoring the loss of the adventure piece that travel had had until then (Bernal, 1985). From that point onwards, the romantic interest in Andalusia turned into visits to the main capitals by train, which were its only stops on its routes (Héran, 1979). Returning travelers complained about finding themselves in a commodified territory, with dances organized for tourists and an increase in prices. In this way, the picturesque that had generated so much attraction for romantic travelers began to be commercialized by its advocates. Most texts and engravings were articulated around topics such as the wild, Middle Ages, or Oriental culture, which were constantly exalted as reminiscences of a distant and extravagant Arab world.

5.3.3 STATE TOURISM PROMOTION DURING THE 20TH CENTURY

The creation of the first administrative bodies in charge of tourism promotion in Spain at the beginning of the 20th century attests to the importance that this sector was beginning to gain in the national economy (Barke & Towner, 1996; Pack, 2008). Among the functions of these institutions was that of effectively conserving the artistic, monumental, and picturesque richness of Spain while promoting the arrival of foreigners (Fernández, 2007). In this way, the beginning of Spanish tourist promotion was based on those symbols, landscapes, and images that had been exalted decades ago. That is the materials through which a visit to the country began to be marketed derived from a constant ambition on the part of national institutions to consolidate the romantic image of the country during a period in which the bourgeois tourist, coming from Europe and with a high cultural level, continued to demand destinations where they could delight in its historical and cultural legacy (López, 2001). Monuments such as the Alhambra, the Mosque of Córdoba, or the Giralda, as well as personalities with popular character traits, were projected as incarnations of myths with strong evocative power.

Much later, the flourishing of travel guides and tourist maps in the early 1950s (Barke & Towner, 1996) reinforced the image of provincial capitals while avoiding rural regions that offered the visitor the image of an underdeveloped territory with infrastructural deficiencies. The Franco dictatorship (1939–1975) prohibited the possibility of contracting a trip to Spain from a foreign agency, which guaranteed the channeling of foreign tourist groups to regions loyal to the regime where the damages from the war were less evident (Pack, 2008). From that point onwards, the country was projected as a "solid preserver of the authenticity of its culture" (Fodor, 1952) or as "one of the few remnants of an almost lost world" (Williamson, 1951). In this way, some of the most attractive cultural practices for visitors, such as bullfights, gypsy dances, or patron saint festivities, became associated with an exotic domestic culture (Barke & Towner, 1996) (Figure 5.2).

FIGURE 5.2 Tourist posters based on the Andalusian idiosyncrasy were published in the first half of the 20th century by the National Tourism Board and the National Tourism Directorate.

Source: Catálogo del Centro de Documentación Turística de España, www.tourspain.es.

Subsequently, the rise of Fordist tourism would displace the primacy of historical monuments in the embodiment of Andalusian territory towards a new model, in which maritime spaces dominated posters, guides, and magazines, and around which state tourism policies began to revolve. Despite this, on a cultural level, the overestimation of the Andalusian image and its manipulation during the dictatorship would maintain it as a representation of everything Spanish, with heritage references such as clothing or architectural details being introduced in the majority of published promotional resources. In democratic times, state campaigns continued to influence romantic approaches to heritage (Lois, 2018), with posters in which monumental resources, flamenco culture, or traditional gastronomy shared prominence with the sun and the beach (Figure 5.3).

FIGURE 5.3 State campaign posters were edited from 1950 to 2004.

Source: Colección de carteles turísticos del Centro de Documentación Turística de España, www.tourspain.es.

5.3.4 COMMUNICATION BY THE REGIONAL GOVERNMENT

In the 1980s, territorial promotional powers were delegated to the autonomous communities. During the years following the distribution of powers between the Central Government and the Autonomous Communities, a specific public entity was created for the development of Andalusia's promotional policy. From then onwards, the campaigns developed in Andalusia would be marked by a continuous trend with state policies (Hernández, 2007, 2008), including the incorporation of a symbolic language and the use of plans that dealt with placing the elements of heritage being used outside of historical time.

Among the objectives of the first campaigns was to reflect a modern Andalusia while remaining faithful to its traditions, using images of the region's natural and cultural heritage, among which its beaches, flamenco culture, white villages, and iconic monuments such as the Alhambra. In addition, warm tones were used, the disappearance of people from the frames, or the projection of images that showed a fragile nature. These elements would enhance the nostalgic tone of the promotion (Hernández, 2007), favoring a timeless presentation of the region.

At the end of the 1990s and during the beginning of the next millennium, the advertising campaigns launched by the Ministry in charge of tourism had the objective of projecting the image of a modern land that would not forget its traditions. This involved constantly using images depicting the cultural and natural heritage of the region, amongst which iconic monuments such as the Alhambra stood out. Faced with this trend, the "Andalusia Loves You" Campaign launched in 2004, proposing a promotion aimed at different segments to diversify the market and reduce seasonality by portraying a more varied destination. However, the presence of goods with Islamic nostalgia in the posters published at that time continued to account for a significant volume of the promotions.

The last two campaigns launched at the regional level, "Your Best You" and "Intensely," also dealt with showing the infinite Andalusian offer, despite repeatedly resorting to images of regional folklore (Cobos, 2015) and architectural sights already common in 19th-century works, such as the Alhambra, the Mosque of Córdoba, or the Alcázar of Seville, as well as the Arab Baths of Jaén, and the Hammam of Córdoba. For authors such as Fuentes (2005), Lasso (2006), and Albarrán & Calle (2021), the latest campaigns published by the Andalusian public administration represent a reiteration of elements typical of the most topical image of the region. Thus, the published material insists on resources such as flamenco, Muslim architecture, gastronomy, or good weather, since their objective is to reinforce the territory brand in international markets (Martínez & Nicolás, 2014). For this, they use representations already consolidated in the general imagination together with the presentation of the target audience itself.

5.4 DISCUSSION AND CONCLUSIONS

In Andalusia, first at the state level and then from the autonomous public management level, the idea of the exotic and the oriental has been largely present in tourist promotion developed by the first public administrations in charge of tourism management from the beginning of the 20th century to today. Frequent in these promotions was the incorporation of textual and pictorial content that had already been spread throughout Europe during the 18th and 19th centuries, and which had their origins with the commencement of interstate travel inspired by the Enlightenment. Then, inside the context of a great flourish of travel literature, visitors extolled the Muslim imprint of a region that was distanced from the hegemony of the rest of the continent to the point of making it the main basis of their international cliché. This was something that strongly influenced the perception that the region would have through its universal image.

Consequently, public administrators have constantly resorted to this image later on; an image that appears biased and based on the merely picturesque, because with that they could achieve a good commercial position in an increasingly competitive market. Thus, in Andalusia, where tourism has transcendental economic importance, official institutions have based the induced image of the region on traits of the romantic ideal. This trend has favored the configuration of an induced tourist image that is distorted and based on the principles of fragmentation – created from small, selected pieces of reality, staging – from its connotative character, romanticization – through the discourse of the sublime and exotic – and stereotyping – seeking to reproduce frozen time through key icons.

This phenomenon has led to a channeling of tourist flows towards those spaces occupying a strong position in the universal imaginary, while other spaces of great historical relevance but that remain distanced from the hegemonic idea of romantic Andalusia are ignored. This process delineates a pattern in market demand, since the most requested elements from the tourism issuing centers, which create the universal image of the territory, are the ones that are best positioned in regional promotion. They represent icons of easy photographic appropriation and of great historical relevance that deeply influences the consumer during their process of choosing a destination for travel. It is in this way, as Dachary & Arnáiz (2009) would affirm, that the myth of paradise has become the fundamental axis of the modern tourism promotion industry. Furthermore, in a context where numbers continue to prevail in the evaluation of tourism success, public administration authorities follow the echo of spreading a fully consolidated, attractive, and alluring image, guaranteeing the seduction of a large volume of travelers despite the risks that this may entail for heritage conservation.

The construction of this identity has been carried out through a series of elements and symbols that, as part of an iconographic heritage, have been consolidated as attributes of a territory using primarily visual discourse. This fact is consistent with what was stated by Troitiño & Troitiño (2011), who argued that today visitors continue to focus their attention on a very limited number of goods and spaces similar to those from centuries ago when tourism was only practiced by the wealthiest social classes.

In conclusion, Andalusian heritage has had a transcendental importance in the construction of the image of Andalusia, largely due to the interest that existed during the Enlightenment for the societies of the past and, therefore, for the testimonies left behind. The research has made it possible to determine that, although the projection made by the entities responsible for tourism promotion is diverse and heterogeneous, stereotypes have not ceased to be a constant. Thus, monumental wealth has continued to be the main protagonist in the campaigns carried out in recent decades, despite the fact that the region has been linked to an increasingly varied tourist offer. These campaigns have been specially led to a large extent by buildings of medieval origin. In the same way that foreign travelers did centuries ago, these have been projected as scenarios where they can live unique experiences (Rodríguez, 2000). And it is that most of the spaces represented in these campaigns suppose the same elements that were already exalted by British, French, or German travelers during their trips through Andalusian lands centuries ago. Therefore, there is a romantic inertia in the projected representations of Andalusia. This continuity of elements from the past would be based on the fact that they represent attractions that are easy to photographically appropriate. It means they represent perceptions well established in the collective imagination, thus facilitating their recognition in a complex and varied market like the current one (Albarrán & Calle, 2022).

Tourism is a sector ruled by the visual, encouraging seductive representations of destinations to attract potential visitors. That is why; in order to achieve a more sustainable and orderly tourism development, it is necessary to carry out an analysis that avoids the mere use of economic data to advance its approaches. Tourism understood as an activity with a strong and transforming power at the spatial level, must respond to a more holistic, transversal, and truthful conception of place. The reproduction of this methodology in other contexts, where tourist imagery also enjoys great international relevance, would assist in detecting the existence of common patterns.

KEYWORDS

- **Andalusia**
- **economic data**
- **heritage**
- **primarily visual discourse**
- **semiotic analysis**
- **tourism promotion**
- **tourist image**

REFERENCES

Albarrán Periáñez, J. D. (2020). Patrimonio territorial, políticas públicas y turismo: un análisis escalar. Los casos de Andalucía y la Sierra de Cádiz. [Tesis]. E-prints complutense. Repositorio institucional de la UCM. https://eprints.ucm.es/id/eprint/64891/.

Albarrán, J. D. (2021). Territorial image construction: An original perspective from the outsider. In M. Gravari-Barbas & I. Stanković (Eds.), *Heritage Hybridisations Concepts, Scales and Spaces. Una Europa PhD* (pp. 105–108). Paris: Una Europa PhD Workshop Paris 1 Panthéon-Sorbonne.

Albarrán, J. D., & Calle, M. D. L. (2021). El patrimonio en la imagen turística inducida de Andalucía. *Cuadernos de Turismo, 48*, 457–490.

Albers, P., & James, W. R. (1988). Travel photography: A methodological approach. *Annals of Tourism Research, 15*(1), 134–158.

Andreu, L., Bigné, J. E., & Cooper, C. (2000). Projected and perceived image of Spain as a tourist destination for British travelers. *Journal of Travel & Tourism Marketing, 9*(4), 47–67.

Barke, M., & Towner, J. (1996). Exploring the history of leisure and tourism in Spain. In *Tourism in Spain: Critical Issues* (pp. 3–34). Wallingford: CAB International.

Barthes, R. (1957). *Mitologías*. Ciudad de México: Siglo XXI Ediciones.

Bernal, M. (1985). *La Andalucía de los libros de viajes del siglo XIX (Antología)*. Seville: Editoriales Andaluzas Unidas.

Bourgoing, J. F. (1808). *Tableau de l'Espagne Moderne*. London: John Stockdale.

Buck, R. (1977). The ubiquitous tourist brochure: Explorations in its intended and unintended uses. *Annals of Tourism Research, 4*(4), 195–207.

Calatrava, J. (2018). Turismo de masas y banalización de la historia: Granada y la Alhambra como caso de estudio. In *Jornadas La Ciudad del Turismo, Apropiaciones Estéticas y Patrimoniales Contemporáneas*. 35–41. Alcalá de Henares: University of Alcalá de Henares.

Carrère, E. (2001). La imagen desértica de Andalucía a través de los relatos de los viajeros románticos franceses: ¿mito o realidad? *Cuadernos de Ilustración y Romanticismo, 9*, 29–46.

Castro, L. P. M., & Rodrigues, J. R. (2017). El uso de la imagen turística en la promoción de los destinos patrimoniales. *Tourism and Hospitality International Journal, 8*(2), 51–67.

Cobos, M. A. (2015). *Publicidad Turística en España* (Workorder degree). Malaga: University ofMalaga: http://hdl.handle.net/10630/10318.

Dachary, A. C., & Arnaiz Burne, S. M. (2009). Colonización, turismo e imaginarios en el siglo XX. In *XXVII Congreso de la Asociación Latinoamericana de Sociología: VIII Jornadas de Sociología de la Universidad de Buenos Aires*. Buenos Aires: Latin American Association of Sociology.

Dann, G. (1996). Tourists' images of a destination: An alternative analysis. *Journal of Travel and Tourism Marketing, 5*(1–2), 41–55.

Dilley, R. S. (1986). Tourist brochures and tourist images. *The Canadian Geographer, 30*(1), 59–65.

Donaire, J. A., & Galí, N. (2011). La imagen turística de Barcelona en la comunidad de Flickr. *Cuadernos de Turismo, 27*, 291–303.

Edelheim, J. R. (2007). Hidden messages: A polysemic reading of tourist brochures. *Journal of Vacation Marketing, 13*, 5–17.

Fernández, J. (2007). *Los Paisajes Andaluces: Hitos y Miradas En Los Siglos XIX y XX*. Seville: Ministry of Culture, Junta de Andalucía.

Fernández, S., Fernández, V., Hernández, E., López, E., Quintero, V., Rodrigo, J. M., & Zarza, D. (2010). *Paisajes y patrimonio cultural en Andalucía. Tiempo, usos e imágenes*. Seville: Andalusian Institute of Historical Heritage.

Fodor. (1952). *Spain and Portugal in 1952*. New York: Fodor Travel Publications.

Fuentes, A. (2005). La traducción de promoción turística institucional: la proyección de la imagen de España. In *La Traducción en el sector Turístico* (pp. 59–92). Granada.

Fuertes-Manjón, R. (2004). El viaje del Mayor W. Dalrymple por tierras leonesas: una visión ilustrada del mundo maragato. *Argutorio, 14*, 51–54.

Galant, I. (2018). Hitos patrimoniales sevillanos: permanencias y novedades en el sector turístico. In *Jornadas La Ciudad del Turismo, Apropiaciones Estéticas y Patrimoniales Contemporáneas*. 59–63.Alcalá de Henares: University of Alcalá de Henares.

Garayo, J. M. (2001). Los espacios naturales protegidos: entre la conservación y el desarrollo. *Lurralde Investigación y Espacio, 24*, 271–293.

García, F. (1995). Descubriendo la ciudad (I). El urbanismo malagueño según los viajeros extranjeros del siglo XIX. *Baetica. Estudios de Arte, Geografía e Historia, 17*, 7–29.

García, J. (1999). *Viajes de extranjeros por España y Portugal: desde los tiempos más remotos hasta comienzos del siglo XX*. Valladolid: Ministry of Education and Culture, Junta de Castilla y León.

Goodrich, J. N. (1978). The relationship between preferences for and perceptions of vacation destinations: Application of a choice model. *Journal of Travel Research, 17*(2), 8–13.

Guerrero, A. C. (1990). *Viajeros británicos en la España del siglo XVIII*. Madrid: Aguilar.

Gunn, C. (1988). *Vacationscape, Designing Tourist Regions*. New York: V.N. Reinhold.

Héran, F. (1979). L'invention de l'Andalousie au XIXe S. dans la littérature de voyage: Origine et fonction sociales de quelques images touristiques. In *Tourisme et développement régional en Andalousie* (p. 29). Paris: Éditions E. de Broccard.

Hernández, J. (2007). La parte por el todo. Imágenes de Andalucía emitidas por organismos públicos. In *La imagen de Andalucía en el discurso turístico*. 26–53. Seville: Factoría de Ideas, Centro de Estudios Andaluces.

Hernández, J. (2008). *La imagen de Andalucía en el turismo*. Seville: Fundación Centro de Estudios Andaluces.

Lasso, C. (2006). *La imagen de Andalucía en el discurso publicitario*. Seville: Centro de Estudios Andaluces.

Lerivray, B. (1975). *Guides Bleus, Guides Vertes et Lunettes Roses*. Paris: Du Chef.

Lois, R. C. (2018). Planificación y gestión territorial del turismo, en Teorías, modelos y destinos turísticos en épocas de cambios. In *XVI Coloquio de Geografía del Turismo, el Ocio y la Recreación de la AGE*. 13–26. León: University of León.

López, A. (1988). El paisaje de Andalucía a través de los viajeros románticos: creación y pervivencia del mito andaluz desde una perspectiva geográfica. In *Viajeros y Paisajes* (pp. 31–65). Madrid: Alianza Editorial.

López, A. (2001). Caracterización geográfica de Andalucía según la literatura viajera de los XVIII y XIX. *Eria, 54–55*, 7–51.

Martín, B. (2003). La imagen turística de las regiones insulares: las islas como paraísos. *Cuadernos de Turismo, 11*, 127–137.

Martínez, E., & Nicolás, M. Á. (2014). The construction of tourist space by public administration and institutional communication: The image of the brand Andalucía as a tourist destination. *Journal of Promotion Management, 20*(2), 181–199.

Matteucci, X., & Önder, I. (2018). Using drawings to explore images of Vienna. *Tourism Analysis, 23*(4), 517–531.

McDonald, J. (2015). The role of ethics on tourist destination image formation: An analysis of the French student travel market. Sofia: Business Administration. Université Nice Sophia Antipolis. Eds.

Méndez, M. (2017). El estilo del relax y la imagen pops de la Costa del Sol. In *Immaginare il Mediterraneo* (pp. 101–110). Naples: Artstudio Paparo.

Milman, A. (2011). The symbolic role of postcards in representing a destination image: The case of Alanya, Turkey. *International Journal of Hospitality and Tourism Administration, 12*(2), 144–173.

Novo, G., Osorio, M., Torres, J., & Esquivel, E. (2013). Viajes, actuantes, escenarios e interacciones: Un análisis de la publicidad de los destinos, a partir de sus semánticas visuales. *Investigaciones Turísticas, 6*, 27–46.

Pack, S. D. (2008). Tourism, modernization, and difference: A twentieth-century Spanish paradigm. *Sport in Society, 11*(6), 657–672.

Perelló, J. L. (2006). La imagen del destino turístico en el proceso global de satisfacción del consumidor: Metodología para su estudio. Habana: ETUR.

Picazo, P., & Moreno-Gil, S. (2017). Analysis of the projected image of tourism destinations on photographs: A literature review to prepare for the future. *Journal of Vacation Marketing, 25*(1), 3–24.

Picón, E., Varela, J., Rial, A., & Braña, T. (2013). Habi: Una herramienta para la representación de la imagen de un producto o servicio. *Revista Española de Investigación de Marketing ESIC, 17*(1), 83–104.

Pritchard, A., & Morgan, N. J. (2001). Culture, identity and tourism representation: Marketing Cymru or Wales? *Tourism Management, 22*(2), 167–179.

Rodríguez, A. J., Díaz, P., & Santana, A. (2012). Estrategias de gestión de imagen de destino en Fuerteventura: De los folletos a la intercomunicación. *Cuadernos de Turismo, 30*, 219–239.

Rodríguez, F. (2000). El paisaje de España y Andalucía en los viajeros románticos: El mito andaluz en la perspectiva geográfica actual. *Proceedings of the II Conference of Hispanists of Russia. Moscow, April 19–23, 1999.* Madrid, Published by the Embassy of Spain in Moscow.

San Martín, H., & Rodríguez, I. A. (2008). Exploring the cognitive-affective nature of destination image and the role of psychological factors in its formation. *Tourism Management, 29*(2), 263–327.

San Martín, H., Rodríguez, I. A., & Vázquez, R. (2006). Análisis de la imagen en turismo mediante técnicas estructuradas y no estructuradas: Implicaciones competitivas para los destinos turísticos. *Revista Asturiana de Economía, 35*, 69–91.

Santillán, V. L. (2010). Fotografía como creadora de la imagen de un destino turístico: Buenos Aires a través de sus tarjetas postales. *Pasos, Revista de Turismo y Patrimonio Cultural, 8*(1), 71–82.

Santos, X. M. (2008). Paisaje e identidad en la promoción de la imagen de los destinos turísticos. In *Espacios Turísticos: Mercantilización, Paisaje e Identidad* (pp. 195–206). Alicante: University of Alicante.

Singh, N., & Lee, M. J. (2009). Convergence and congruency of pictorial destination images in DMOs' websites and brochures. *Journal of Hospitality Marketing and Management, 18*(8), 845–858.

Soler, E. (2013). Imagen de la moderna España. Alicante: University of Alicante.

Thurot, J., & Thurot, G. (1983). The ideology of class and tourism: Confronting the discourse of advertising. *Annals of Tourism Research, 10*(1), 173–189.

Torres, J. (2017). Los premios de embellecimiento: Paisaje e identidad turística local. In *XXV Congreso de Geógrafos Españoles: Naturaleza, Territorio Y Cultura En Un Mundo Global.* (pp. 2309–2318). Madrid.

Troitiño, M. Á., & Troitiño, L. (2011). Estudio sobre la funcionalidad turística del patrimonio cultural del municipio de Carmona y estrategias de actuación. *PH Boletín del Instituto Andaluz del Patrimonio Histórico, 78*, 4–13.

Troncoso, C. A. (2013). Política turística y patrimonial en la ciudad de Salta, Argentina: Disonancias en la protección y usos del centro histórico. *Estudios y perspectivas en turismo, 22*, 1002–1024.

Urry, J. (1990). *The Tourist Gazes*. London: Sage.

Urry, J. (2002). *The Tourist Gaze: Leisure and Travel in Contemporary Societies*. London: Sage.

Williamson, H. W. S. (1951). *The Tourist Guide-Book of Spain*. London: The Times of Spain.

Woodside, A. G., & Lysonski, S. (1989). A general model of traveler destination choice. *Journal of Travel Research, 27*(4), 8–14.

CHAPTER 6

TĪRTHA DARŚANA: PILGRIMAGE-CENTRIC CONSTRUCT AND RECONSTRUCT OF VARANASI/Kāśī

KRISHNA KUMAR,[1] SHANKHA PRATIM BHATTACHARYA,[2] and PULAK MISHRA[3]

[1]*Research Scholar, Center of Excellence for Indian Knowledge Systems, IIT, Kharagpur, West Bengal, India*

[2]*Associate Professor, Architecture and Regional Planning, IIT, Kharagpur, West Bengal, India*

[3]*Professor (Economics), Department of Humanities and Social Sciences, IIT, Kharagpur, West Bengal, India*

ABSTRACT

'Pilgrimage' can be considered the earliest mention of 'Tourism' in all civilizations. The earliest mentions of Hindu Pilgrimage (*Tīrthayātrā*) can be sourced in ancient epics, mythologies, and regional folklore. These can be traced back as far as the fifth century BCE, and even further back into the chronicles of human civilization. These works of religion very well describe the motivation and merit to visit sacred places. These holy locations and their sacrosanct points were praised in medieval myths. Varanasi/*Kāśī*/Benares is believed to be the oldest holy city on earth. While the early references of these *Tīrtha*-centric texts depict *Kāśī* as a place of ascetics and yogis (5th to 8th centuries) with only a few references to temples and pilgrims, a significant discrepancy can be observed in later additions to these texts. These additions to *mahātmya* texts around the 11th to 15th century AD have spurred the

Dynamics of Heritage Management and Tourism Development: Trends, Practices, and Approaches.
Surabhi Srivastava & Ruchika Kulshrestha (Eds.)

devotion and pilgrimage towards 'glorifying' *Kāśī.* The cognitive image of the city as instilled in the mind of visitors with the feeling of self-fulfillment still has a lot to offer to every visitor. The main framework of Sustainable Management of Pilgrimage and Heritage tourism to Varanasi considers religion and religious literature as a perspective of usefulness for its survival as an individual or group. The present research focuses on observing the dynamics of literary references in promoting pilgrimage-centric religious Tourism to the sacred territory of Varanasi/*Kāśī.* Such approaches to traditional observations of Sustainable Management practices of religious centers and sacred places will guide us to build a strategic framework for specialized, sensitive, and sustainable management visions toward Heritage Tourism and Urban Conservation of Historic Indian Cities.

6.1 INTRODUCTION

Pilgrimage, as considered in the Indian subcontinent, is the oldest form of the motivated journey undertaken to a religious and sacred place (Singh, 2009), *Tīrtha*. sacred, the divine act of nature, paths, riverbanks, and sacred landscapes, regarded as 'Pratyaksh dev' – 'Prima facie of Divine,' and the journey so traveled is considered *Tīrtha* yātrā (Bhardwaj, 1973; Bakker, 1990). The essence of this faithscape is glorified in the epigraphs of the traditional knowledge systems and composed as Tirtha mahatma and the sthalpuranas. Pilgrimage to such *Tīrtha* destinations has been a widespread practice since the Vedic ages, as mentioned in Rigveda[1] (Bharati, 1963). While such travels were a difficult task with limited resources, their purpose and benefits established their motivation. Pilgrimage, trade, or journey, amusement/tourism, can have multiple names based on their purpose of movement (Coleman, 1995). However, such connotations were not as sharp in their meanings as they are in the present day (Dubey, 2000). Around the time the Puranas were written, the marketing of locations based on their *Tīrtha* status rose to the forefront.

Varanasi is the oldest among the world settlements that have survived to the present day. The city is an established seat of a supreme *Tīrtha,* as glorified in these epigraphical references (Eck, 1986). The city is observed as a confluence of multi-diversified cultures, religions, and philosophies. Scholars of Indian culture and traditions have observed Varanasi as the greatest

[1]The mentions of the pious Sarasvati River, along which Vedic hymns were composed. *Taittiriya Aranyaka* describes about the Mansarovar at Kailash, Kāśī, and another *tirtha.*

veneration of antiquity and continuity in the world (Singh, 1988). The city has been considered the supreme seat of pilgrimage for Hindus, the supreme *Tīrtha* as scholars have eulogized, with archaeological and textual evidence that claims the continuous construction and reconstruction of *Kāśī* (Singh, 2019). Described as *Aavimukteśvara* (never forsaken), or *Ānaṃdakannana* (forest of bliss), or the eternal city or the city of lights, its mentions in Vedic literature reiterate the importance of *Kāśī* for offering spiritual and material benefits (Sukul, 2000). The *Kāśī Khand* of *skaṃdapurāṇ*a describes the city's capability to offer moksha (eternal liberation) from the material world. The great epics of India have illustrated the city as Tirthasthal, a sacred place of religious importance, called *Kāśī* kshetra (Singh, 1993). Recent archaeological and textual sources have testified to its construction in the early millennia BCE (Jaiswal, 2009). Located at the junction of two rivers, the river Ganga and Varuna (also mentioned as Varanasi River), the city brings wealth and development as an important commercial and trade center. Owing to this strategy, the city attracts various groups of people (Kutlutürk, 2013). The established historical settlement of Sarnath (a satellite settlement, 5 km north of the present city) dating back to the 5th century BCE testifies to the urban expansion and scale of the city in its earliest days (Singh, 1993). However, the claims of 'holiness' in the archaeological and textual testimonies had special significance for sacred and religious attainment within the Brahmanical mainland. In the further centuries of history, the popularization of this faithscape led to transformations and gradual constructs and reconstructs of the city (Bhardwaj, 2009). Such transitions of sacred and profane construct the heritage of the present. The transformation of trade and commerce, i.e., profane space into sacred space (*Tirtha*), could incidentally support its survival and sustenance. This was further reinforced under the patronage of the Gupta and Gahadavala dynasties (Bharati, 1970). The so-developed impression of *Tīrtha* invited various aspects of sacredness and spirituality to *Kāśī*, which emerged as different disciplines of spirituality, religion, and philosophies (Feldhaus, 2003). This is observed to be the peculiar aspect of the development of *Kāśī* into Varanasi, the oldest surviving Heritage city in the world. An aspect of the city that would be appreciated is its growth and prosperity with sacredness and holiness (Glushkova, 2003). Hence, the town has much to offer to ascetics and yogis, to art and craft industries, to music and literature, and to social and political knowledge systems of life. Such constructed imagery of the city has fueled the survival and sustenance of the city on the cognitive plane of the people through ages and generations. Such maintenance, management, and development strategies as practiced in ancient times can lead us to better sustainable future development goals in similar historic Indian cities (Granoff, 2004). The present research is an attempt to bring a sensitive autonomous approach towards development in *Tirth*-centric historic Indian cities (focusing on Varanasi).

6.2 HISTORICITY; *K*āśī FROM PRESENT TO PRESENCE OF PAST

It is believed that the historical city of *Kāśī* was founded in the region that is now the northern portion of the present-day city, specifically on the Rajghat hilltop. The excavations here date the antiquity of the city way back to the ninth century BCE (Jaiswal, 2009). The early references to traditional practice in *Kāśī* are observed to be very similar to those in the nearby hinterlands. The ascetic and cultic practices observed in the region were a very primitive form of nature worship (Sukul, 2000). A recent investigation from Vedic literature has identified *Kāśī* as an important pilgrimage destination, *KāśīTīrtha* darśana. *Manusmiriti* mentions *Kāśī* as part of the Āryāvarta (Saraswati, 1985). This further got equalized in the Puranic texts, where *Kāśī* emerged as an important *Tīrtha*. While the ancient epic Mahabharata mentions *Kāśī* as the supreme *Tīrtha* (Singh, 2002), the conflict that took place at Kurukshetra, *Kāśī* is presented as one of the significant countries that existed during that time (Malville, 2004). Though the geographical area of *Kāśī* could have been centered around the Rajghat, the recent excavations bring new insights into its settlements and expansions up to nearby hinterlands.

The visit of Lord Buddha to *Kāśī* has brought a lot of urban and religious changes to the city (Saraswati, 1975). With various missionaries and ideologies traveling to the city, the worldwide expansion of Buddhism and the construction of pillars, stupas, and temples at Sarnath in *Kāśī* were all observed by Emperor Ashoka. Although *Kāśī* was the birthplace of Jain Tirthankaras, there is not much clear evidence of its spread in regions and neighborhoods (Singh, 1955). A new traditional form of built archetype was introduced to the city as Buddhist Viharas, monasteries, and stupas around 500–200 BCE. This also brought an organized new social structure of Sangha (Nanamoli, 1972). Sangha, the Buddhist pattern of organized designations in religion as monks, nuns, and followers. New religious mobs floating in town, attract new political administrations and patronages that resulted in the creation of new novel architecture built around which is our Heritage of the present. Despite being a growing new religion, tradition, and practice, *Kāśī* has always remained an important pilgrimage/*Tīrtha* center.

Further, with the growing powers of Shungas and Guptas over central and northern India, *Kāśī* has been brought to specific attention (Chakravarti, 2010). This led to the construction of magnificent temples in the sacred territory of *Kāśī*. Recent excavations have uncovered the remains of temples in the nearby areas of Varanasi, dating back to the period of Skand Gupta (Shobana, 2011). It is believed that during the reign of Vainya Gupta, approximately from 500 to 508 AD, the main shrine of *Kāśī* Vishwanath was built. This is thought to have taken place around the same time that the city's religious foundations were strengthening. Historians and scholars have noted the accounts of the Chinese traveler Xuanzang describing various temples in the town (Motichandra, 1962), urban structures, gates, and markets. The progressive spread of Shaivism and Advaita philosophy in India through the movements of Adi Shankaracharya (around 600 AD) unified various cultic and religious practices, forming a larger network of *Tīrtha*s (Singh, 2009). The presence of the Jangambarimuth in *Kāśī* is evidence of the close connections between that city and the Deccan region, as well as the widespread adoption of Shaivism across the Indian subcontinent. This further increased the influx of pilgrims and the development of urban landscapes in such sacred cities. The strategic location of the town with access to major trade routes, combined with religious networking as *Tīrtha*s, ushered in the golden ages of urban *Kāśī*.

During the reign of Islam in India around 1194 to 1700 AD, the renowned Hindu pilgrimage of the country faced much destruction (Singh, 2009). The city has witnessed a series of attacks and massive destruction in the name of religion (Motichandra, 1962). This episode of damage intimates us about the wealth of the city (built tangible archetypes and intangible religious beliefs and traditions). Also, this period brought us other constructive skills in the form of art and designs (Sherring, 1868). The so-migrated Muslim settlements have started settling at different places and the city tends to expand towards its southern edge. The destruction of physical constructions was a direct consequence of the political confrontations that took place (Eck, 1982). The conflict that was raised out of a religious core was again then revived/repaired/reconstructed out of religious beliefs only. However, this brought a new style of architecture to the town (Gutschow, 1994). The construction of Indo-Islamic structures, with mosques, domes, and tall minarets. While Raja Todar Mal built the temple for *Kāśī* Vishvanath in 1585 under the rule of Akbar. The Jantar Mantar and Observatory at Dashashwamedhghat were also built around the same time period by Raja Man Singh Ji. The layer of tangible structures that we see at present days is the only surviving construct of this reconstructing phase of the city.

With the growth and popularity of religious beliefs in the country, Maratha rulers took an interest in *Kāśī* (Motichandra, 1962). With the growth of their political power in the region, *Kāśī* received a lot of attention. Shrimant Baji Rao Peshwa-I commissioned the construction of the present temple of *Kāśī* Annapurna around 1725 (Havell, 1905). After the demolition of the *Kāśī* Vishvanath temple by Aurangzeb, Maratha queen Rani Ahilya Bai Holkar united all Hindus and built the present Temple of *Kāśī* Vishvanath in 1775. The settlement of Maratha around the Panchgangamohalla of present Varanasi can still narrate the stories of their migration (Lannoy, 2002). This promoted the building of ghats along the bank of the river Ganga, new languages and customs also started to take root in the culture and traditional practices of the town, hence making it a pluralistic society, an impression of mini-India – *Akhaṃḍa Bhārata* (Eck, 1998). As the British Empire started spreading its influence in the country, the urban landscape of the city also began to change. During this time, the city began to see an influx of colonial and Gothic-style architecture (Singh, 2002). The mint house, cantonment, water pumping station, and main building of Sampurnand Sanskrit University are a few such British establishments in town. This is the time when Sir James Princep prepared the first cartographic map of Varanasi (Singh, 2013). James Princep deserves a lot of credit for everything he's done for Varanasi. The city's population was estimated to be around 2 lakhs as of the 1821 census. The increase in urban structure and the progressive construction works and trade had attracted various population inflow into the town. The floating population is going to be increasing in the town with better transport connectivity (Sukul, 1974). This was the time when scholars from the West explored *Kāśī* from its most accepted pilgrimage part. Whereas Mark Twain noted that the city was more ancient than History, older than Tradition, and older than Legend, and it appeared to be twice as old as all of these things combined.

6.3 *K*āśī AS TIRTHA

Kāśī/Varanasi is the most celebrated pilgrimage city for Hindus (Vidyarthi, 1979). Its spiritual manifestation has not just possessed all the divine representations in the city but also formed Pan-Indian imagery of art and architecture in its urban framework. As described in the *Kāśī*Khand of *SkandPuran* (Eck,

1998), all the other *Tīrtha* and their respective deities have their residence in *Kāśī*, out of which we see all sacred places of Hinduism in abbreviated form and specially transported into the cultural landscape and sacred geography of the city. That is how scholars have identified the city as the microcosmic representation of the macro cosmogram (Humes, 1993). Also, it is considered the territory of an imagined landscape (Eck, 1986). Being seated as the supreme *Tīrtha* in Hinduism, *Kāśī* has been the center for the vast collection of knowledge systems that are well reflected in sacred literature and mythologies. Located at the western banks of the northern flowing Ganga, *Kāśī* is recognized as an ancient *Tīrtha,* a crossing-over where the material meets spirituality (Badone, 2004). According to the mythological stories, these *Tīrtha* places are doors to spirituality where the divine gets access to the earth and human souls can meet the divine after liberation (Bhardwaj, 2004). Among all the other *Tīrtha* places mentioned, along with Prayagraj and Gaya, *Kāśī*/Varanasi is among the greatest *Tīrtha* of India.

The word *Tīrtha* is referred to as the 'ford' or crossing place (Cohen, 1992). Such places of transition were believed to be blessed with spiritual powers. These were initially the places where rivers or the water is shallow in rivers where travelers can rest, drink, and bathe (Caplan, 1997). Such serene calmness brings relief to the body and the soul, hence bringing spiritual peace. In Hinduism, taking a bath is a ritual with profound spiritual meaning (Dubey, 2000); almost all the scales of *Tīrtha* are denoted with a spot of water or water body nearby, where having a bath was a ritual that was highly valued. Varanasi is such a particularly significant *Tīrtha* that would have been determined by the aggregation of historical occurrences affiliated with various travelers, trade, merchants, priests, and the divine. The strategic location of *Kāśī* provided the best of such a suited position (Finlay, 2011). By no coincidence, the main trade route between Calcutta to Delhi and beyond passes through Varanasi. The whole evolved layer is built upon the *Tīrtha*-centric imagery that exists in the cognitive memory of its people. The built-up layers over these millennia were layered upon the memory of its users and that would have sanctified the spot.

Lord Shiva, the supreme deity of *Kāśī* worshipped as *Kāśī* Vishwanath, was a mountain-descended ascetic who needed to locate a suitable residence for his newlywed wife Parvati (Bakker, 1996). Devi Parvati, who is the daughter of the Himalayan King, lived with royal wealth and pleasure. In the search, he scanned the entire earth and selected the location of Varanasi. The beautiful city of *Kāśī*/Varanasi as described is the most beautiful place on earth, with streams, springs, extensive forests, and the free-flowing

Ganga River, adorned with palaces, gardens, and temples. Shiva, when descended from Mount Kailash, became both a family man (from *sanyas* to *grihastha*) and a city dweller in the blissful forest (*Anand Kannan*) (Singh, 1993). Presently, it is difficult to picture how beautiful the city must have been in its heyday, but not very many centuries ago, the city was covered in trees and had waterways, ponds, lakes, and flower gardens throughout. Ashrams and temples were located on the banks of these water bodies and the city was teeming with that image of the imagined landscape so portrayed (Gothóni, 2010). Where the landscape pictured has lived within the cognitive memories of its people. Beyond the edge of the city, beyond the walls of the façade of the ghats, beyond narrow lanes is the network of pilgrimage paths, pilgrimage journeys, and the *Kāśī* of the sacred and cosmic realm (Singh, 2002). Therein lies a complex pilgrimage landscape composed of pilgrim imprints, sanctuaries, temples, and memory, with space-time alignments and multilayered temporal structures.

6.4 CONSTRUCTION AND RECONSTRUCTION: PATTERNS OF TIME

Varanasi is the oldest among the world's living settlements with a three-century-old layer of history. Located at the heart of the present Indian state of Uttar Pradesh and situated along the western banks of the river Ganga. The historic city has been widely acknowledged as the supreme destination for pilgrimage tourism in India. The emerging identities and the emergent requirements have marketed the city as a heritage city (Desai, 2017). The layers of heritage are built upon the base of spiritual/religious cognitive imagery, which is identified in temples, rituals, riverfronts, monasteries, routes, festivals, language, traditions, etc. The city has been recognized as the sacred geography of the present urbanscape as described in the *Tīrtha*-centric pilgrimage literature (Gesler, 2000). While scholars and academicians have observed the city based on an imagined landscape, a cosmic sacred setting in pilgrimage and trade, experiencing a modern metropolis that gives the idea of being timeless (Eck, 1986). Whereas Varanasi has been built, destroyed, and reconstructed up to the boundaries that we can see in the remnants of architecture in the city today (Rana, 2019; Bakker, 1996). The observable shreds of the past that we get to see in the present are the only remains of this repetitive construct and reconstruction. The present layer of Architectural Heritage that is visible on the surface is a result of the invasions and counter-invasions. Its effects can be seen in the forms of the city-built model, which has aspects of Islamic, Hindu, Buddhist, and Colonial architecture. The literary texts, cartographic maps, and symbolic representations of the sacred geography that were tooled to promote the city were mainly about space and played a very important role. The composers of such strategic tools were well equipped with the significance of the sacred sites, location, and physical disposition layered with its continued reinterpretations, and portrayed as the evolving sacredscape of the city. This draws our attention toward the dynamic relationship between religious imagination and its contingent political and cultural contexts (Logan, 2010). This consistently evolved spiritual picturesque over politico-economic benefits have made a deep impression on the cognitive imagery of its users. The intangible and tangible built environment of the city is an archive of its rich and revealing spatial history (Kumar, 2010). Visitors/tourists in present-day experience the city as a living cultural museum. The current urban experience of Varanasi has been formed by public community places and organizations that are inextricably tied to the sacrosanct religious character of the community. However, the new buildings that are introduced to the city's urbanscape are being overlaid over sacred geography and the Puranic writings provide a concise summary of the metaphorical connections that were spread through practice (Desai, 2017).

Hence, the prolonged history of the city can be indicative of its paradigm shift through time. Yet, the city never tends to resist any phenomenal change confronted in its development. Such contestations are being faced in present times too, where the city is struggling to retain its chosen past, the observable shreds of evidence of its origin, evolution, and sustenance. The development trend of *Kāśī* and similar historic Indian cities are also finding their sustainable development ways through general smart-city-like transformations whose primary objective is city-imagining (Singh, 2017). Wherein, the Indian Smart City Mission focuses mainly on the planning of unplanned urban areas. It intends to decongest the dense urban crowd and make space within them for public facilities and basic amenities. Such urban proposals intend to follow an area-based approach, which might work well for more geographically disciplined spaces (Smith, 1984) but organically evolved Heritage cities like Varanasi require autonomic, keen, and sensitive attention toward their sustainable future development.

While the Urban development initiative has been framed at broader levels to cater to multiple cities, it is also very important to understand the autonomy of the Historic Indian Cities. However, Indian pilgrimage towns have more diverse realities and specific problems, like the city of Varanasi which has emerged with a subconscious sense of contestation, are more intended towards the retention of its centuries-old tangible and intangible Heritage, tradition, and ways of life.

6.5 Kāśī REIMAGINED: TOWARDS SUSTAINABLE FUTURE

Heritage has been considered a primary potential and also a challenge in urban growth and development. Varanasi is also facing such challenges where the heritage, which was the inception for this development, was the result of its sacred religious imagery, recognized as the sacred geography and imagined landscape, is now considered a threat in the path of its development. Heritage-centered urban development was initiated with the goal of making Varanasi an economically and culturally vibrant tourist destination (Singh, 2009). But this vision is itself a paradox as the most traditionally rich heritage clusters of Varanasi are exact places where most of the urban problems are concentrated. The city's regular growth tends to divert development to other areas of the city or to replace the existing ones with new ones (Saha, 2022). The development thus happening is either emerging as privatized urban spaces, which shrink public spaces and green covers and are devoid of heritage. Or, the old structures of the historic core of the city are being modified or demolished in the name of development. The architectural heritage of the city is threatened by new materials and constructions. Recent events and construction works in the old portion of the city have demonstrated that even stakeholders of such heritage properties have preferred to rebuild rather than renovate (Rana, 2019). These hence mislead the whole development movement in the cities of heritage and pilgrimage importance where heritage is being threatened by the risk of demolition.

With the promotion of tourism, it is very much required to see culture as the mirror of the evolution of history and tradition. While heritage resource management considers both the protection and maintenance of sacred sites (Prorok, 2003), it is also equally essential to understand that these sites had evolved and survived with the continuous pilgrim movements and ceremonies strata of ancient human interactions with the earth and its mystic power. To conserve and restore such sacred sites, it is crucial to encourage the rediscovery of their neglected cultural heritage, traditions, and practices, which have evolved out of reverence and harmony with the earth as a source and sustainer of life. A constructive and sensitive look is required where it is essential to process a sensitive investigation in search of the origins of place as a fundamental religious impulse (Zaidman, 2001). It is also required to consider the mental construct of these mystical traditions in the cognitive imagery of its users. Understanding cultural systems in both ways, intrinsic and extrinsic: insider and outsider, from the perspective of a human science paradigm, necessitates a holistic understanding of the sacred and heritage for development, as it attempts to reveal the 'whole' of the culture, human psyche, and functions operating there.

In this study, we have observed the self-organized display of sacred pilgrimage and tourism from ancient times. We believe, however, that the human processes of evolution and sustenance have led to the location of the sacred sanctuaries and pilgrimage, which stimulate socio-political and economic sustainability. It is essential to understand that these sacred landscapes have been carved out by the feet of pilgrims' movements that have provided meaning to the entire landscape. The consistent behavior of millions of people who collectively or individually desire a divine or sacred experience in these locations necessitates a sensitive approach to their future development.

6.6 SUSTAINABLE Kāśī: KARDAMESHWAR MAHADEV TEMPLE

Varanasi, as known to all, is among the oldest living cities in the world, even regarded as older than history. Situated along the banks of the Ganga, the city represents a unique case of harmony with the holy river. Manifestation of this is evident in its rich cultural heritage, both tangible and intangible. Varanasi is jeweled with many historic structures like temples, kunds, Dharamshalas, mosques, etc., but very little of it actually exists from the pre-Mughal era; Kardameshwar Mahadev Temple is one such temple.

Kardameshwar Mahadev Temple is the oldest temple in Varanasi. Built-in the 10th century BC under the regime of Gurjara-Pratiharas (Singh, 2018). Some scholars say the temple has its origin in the 2nd–3rd century, by witnessing images of pre-Aryan deities like Yaksha, Naga, Gandharva, and Kubera. These were observed to be worshipped during pre-Aryan to Maurya and Buddhist time periods in cities and regions (Sherring, 1868). Kardameshwar Mahadev Temple has all such layers of history superimposed over it and has seen all transformations during different phases of time. The layers are solely sustained and survived in their setting, location, and communities settled around, which used to frame its self-sustainable temple management.

Kardameshwar Mahadev has religious and mythological associations in the cosmic mandala of Varanasi, i.e., *Kāśī* Mandala/*Kāśī* Kshetra, the sacred territory of Varanasi, along which many pilgrimage yātrās are performed (KR, 1957) (Figure 6.1). These yātrās include Chaurasikos Yātrā, Panchkroshi Yātrā, Nagar Pradakshina, Avimukteshwara Yātrā, and Antargrihi Yātrā, among which Panchkroshi Yātrā is the supreme yātrā destination for Hindu pilgrims (Singh, 2002).

FIGURE 6.1 Kardameshwar Mahadev Temple and Kardam Kund (*Source:* Author).

Kāśī Panchkroshi yātrā is a sacred pilgrimage route along the peripheral edge of the sacred territory of Varanasi (*Kāśī* Kshetra). It is a total of 83.5 km long yātrā, which has to be covered on foot by pilgrims (Goyal, 2016). This yātrā is covered in six days and five nights. There are five major halts/destinations where pilgrims stay overnight. These destinations are Kardameshwar, Bheemchandi, Rameshwaram, Shivpur, and Kapildhara. Each destination has a main temple along with a Kund/water body and a series of Dharamshala for pilgrims to stay.

Kardameshwar Mahadev temple is the first half of the Panchkroshi Yātrā, and it is an important destination for pilgrims from various parts of the region and India.

Kardameshwar Mahadev temple and its setting at the southern edge of the territory of Varanasi are very significant. The area is a significant low-lying area, whose catchment might feed into one of the important rivers in

Varanasi, i.e., the Assi River (Motichand, 1985). Hence, the region has too many surface tanks/kund/Pokhara which are connected by water streams and channels. Kardam Kunda and Kandawa Pokhara are major water structures in the area (Tripathi, 1962–63).

Local communities living in the area are directly dependent on the temple and its setting. The major communities in the area are Kumhar (Prajapati), Gujjars/Yadavs, Brahmins, and Mali. Prajapati is an important community in the area and has a major stake in temple rituals and temple activity (Kumar, 2018). They are among the primary stakeholders. Prajapati gets the basic raw material, i.e., mud from the bed of these water bodies for their livelihood. Yadavs get the grazing area to feed their cattle, and they supply cow dung cakes (gohari) to the potters to bake their pottery. Hence, there is a significant cross-interdependent relationship between the temple, its historic evolution, its setting, and the local community around an interrelated sustainable management practice. At Kardameshwara Mahadev temple, the interrelated relation of tradition, culture, and community has been observed, which was a self-sustainable survival model for temple management. The significant manifestation of the landscape was serving as a local resource for the survival of the communities around. The gradual development of the Temple and precision was supporting the interrelated connection with the temple, community, and landscape. Through these evolutionary processes, the temple has been growing in the cognitive image of the communities, pilgrims, and landscape. But with insensitive modern development, new alarming issues are appearing. These issues are arising for both culture, community, craft, and planning. Attachment to place and values are still alive in its sociocultural activities and other factors interrelated to that. These factors function through regularity, integrity, and constant making and remaking for a sustainable future. Kardameshvara temple and environs possess all such attributes, processes, and historical continuity of maintaining the aliveness of cultural infrastructure, yet the communities, their craft, landscape, setting, and urban character are being affected. Being a living repository of many cultural ensembles and heritages (tangible and intangible) and attributes of cultural landscapes (pilgrimage activities and sacred shapes), Kardameshvara temple and environs could be taken as a model area where the cultural economy can be strengthened through temple and community promotions. Temples and communities can be more seen as a repository of skill, knowledge, and wisdom, rather than a tourist piece of attraction or creative economic proposals. The grounded solutions to new emergent issues shall be observed in their facts of evolution, survival, and sustainability. Hence, it is very important to have a cosmic vision towards global humanism and self-realization.

At Kardameshwara Mahadev temple, the interrelated relation of tradition, culture, and community has been observed, which was a self-sustainable survival model for temple management. The significant manifestation of the landscape was serving as a local resource for the survival of the communities around. With the gradual development of the temple and practices supporting the interrelated connection with the temple, community, and landscape. Through this evolutionary process, the temple has been growing in the cognitive image of the communities, pilgrims, and landscape. However, with insensitive modern development, new alarming issues are appearing. These issues are arising for both culture, community, craft, and planning. Attachment to place and values are still alive in its socio-cultural activities and other factors interrelated to that. These factors function through regularity, integrity, and constant making and remaking for a sustainable future.

6.7 CONCLUSION TOWARDS REALIZING A SUSTAINABILITY PERSPECTIVE

The composition of these *Tīrtha* mahatmas and Purāṇic literature has long authorized these pilgrimage centers as an interplay of local and world views. The role of spiritual merits and benefits that Puranic literature portrays play an important role in transforming these pilgrimage centers into Heritage cities of the present. The contribution of Puranic literature has opened a variety of possibilities to explore the ways of sustainable urban management and tourism framework. The motivation of sanctification, innovating and commanding pilgrimage, social concerns, socioeconomic needs, and political identities, all supported the array of purposes that pilgrimage serves. This research shows that Historic Indian cities, such as Varanasi, are not just for pilgrims, tourists, travelers, or locals, but they serve all who come into contact with them and have influenced them all; hence the sustainable development goals in such regions must also look into it with the same sensitivity. Religious heritage has to be seen as a resource for sustainable development in Historic Indian cities. In order to productively utilize heritage resources in such pilgrimage cities, first, it is needed to document the resources, extensive background and understanding of these heritage, and its interrelated ancient sustainable heritage patterns, and then protect it for further optimizing it under heritage guidelines and legislations, within the purview of religious-cultural images.

Then, with these effects, gradual awareness can be spread among the locals, users, and stakeholders for their participation, then possibly the interventions can become sustainable – environmentally, socially, and culturally.

KEYWORDS

- **conservation**
- **heritage guidelines**
- **heritage tourism**
- **historic Indian cities**
- **pilgrimage**
- **religious tourism**
- **sustainable management**

REFERENCES

Badone, E. A. (2004). Approaches to the Anthropology of Pilgrimage and Tourisim. In E. A. Badone, *Intersecting Journeys: The Anthropology of Pilgrimage and Tourism* (pp. 1–23). Urbana and Chicago: University of Illinois Press.

Bakker, H. (1996). Construction and reconstruction of sacred space in Vārāṇasī. In *Brill*, pp. 32–55.

Bharati, A. (1963). Pilgrimage in the Indian tradition. *History of Religions*, pp. 135–167.

Bharati, A. (1970). Pilgrimage sites and Indian civilization. In J. W. Elder, *Chapters in Indian Civilization* (pp. 83–125). Dubuque: Kendall/Hunt Publ.

Bhardwaj, S. M. (1973). Sacred Places according to puran and some later sources. In S. M. Bhardwaj, *Hindu Places of Pilgrimage in India: A Study in Cultural Geography* (pp. 58–79). Berkeley, Los Angeles, London: University of California Press.

Bhardwaj, S. M. (2004). Tirtha. In S. Mittal & G. Thursby (Eds.), *The Hindu World* (pp. 478–501). New York: Routledge.

Bhardwaj, S. M. (2009). Circulation and circumambulation. In J. M. Malville & S. M. Bhardwaj (Eds.), *Pilgrimage: Sacred Landscape and Self-Organized Complexity* (pp. 47–58). New Delhi: D.K. Printworld.

Caplan, A. (1997). The role of pilgrimage priests in perpetuating spatial organization within Hinduism. In R. Stoddard & A. Morinis (Eds.), *Sacred Places, Sacred Spaces* (pp. 209–233). Baton Rouge: Geoscience Publ.

Chakravarti, R. (2010). *Exploring Early India: Up to AD 1300*. New Delhi: Macmillan.

Cohen, E. (1992). Pilgrimage centers: Concentric and excentric. *Annals of Tourism Research, 19*(1), 33–50.

Coleman, S., & Elsner, J. (1995). *Pilgrimage Past and Present in The World Religions*. Cambridge, MA: Harvard University Press.

Desai, M. (2017). *Banaras Reconstructed: Architecture and Sacred Space in a Hindu Holy City*. Hyderabad: Orient Blackswan Private Limited.

Dubey, D. P. (2000). *Pilgrimage Studies: The Power of Sacred Places* (Pilgrimage Studies No. 5). Allahabad: Society of Pilgrimage Studies.

Eck, D. L. (1982). *Banaras, City of Light*. New York: Alfred Knopf.

Eck, D. L. (1986). Banaras: Cosmos and paradise in the Hindu imagination. In V. Das (Ed.), *The Word and The World* (pp. 41–55). New Delhi: Sage Publ.

Eck, D. L. (1998). The imagined landscape: Patterns in the construction of Hindu sacred geography. *Contributions to Indian Sociology, 32*(2), 165–195.

Feldhaus, A. (1990). Response to the paper of Hans Bakker. In H. T. Bakker, *The History of Sacred Places in India as Reflected in Traditional Literature: Papers on Pilgrimage in South Asia* (pp. 86–89). Leiden: E. J. Brill.

Finlay, V. (2011). *Green Pilgrim Cities Network: Leaving a Positive Footprint on The Earth*. Bath, UK: ARC (Alliance of Religions and Conservation).

Gesler, W. M. (2000). Hindu Varanasi. *Geographical Review, 90*(2), 222–237. American Geographical Society.

Glushkova, I. (2003). Moving God(s)ward: The idea of Tirtha-Yatra. *Bulletin of the Deccan College Post-Graduate & Research Institute (Pune): Professor Ashok R. Kelkar Felicitation Volume, 62–63*, 256–283.

Gothóni, R. (2010). Pilgrimage: Three theoretical interpretations. In R. Gothóni (Ed.), *Pilgrims and Travelers in Search of The Holy* (pp. 11–30). Oxford and Bern: Peter Lang Publs.

Goyal, A. (2016). Panch Kroshi Yatra of Kāśī – Ancient Pilgrim Route. Retrieved from Inditales: https://www.inditales.com/panch-kroshi-yatra-kashi-varanasi/.

Granoff, P. (2004). Images and Their Ritual use in Medieval India: Hesitations and Contradictions. In P. A. Granoff, *Images in Asian Religions: Texts and Contexts* (pp. 19–55). Vancouver: University of British Columbia Press. doi: https://doi.org/10.59962/9780774851107

Gutschow, N. A. (1994). *Varanasi /Banaras: The Centre of Hinduism. Erdkunde (Bonn),* Bd 48 (3), pp. 194–209.

Havell, E. (1905). The rise of Modern Hinduism. In E. Havell, Benares. *The Sacred City* (pp. 58–70). London: Blackie and Sons Ltd.

Jaiswal, V. (2009). An Overview of Ancient Varanasi. In V. Jaiswal, *Ancient Varanasi: An Archaeological Perspective (excavations at Aktha)* (p. 180). New Delhi: Aryan Books International.

Kāshi Rahasya (1957). *Kāshi Rahasya (an appendix to the Brahmavaivarta Purāna),* No. XIV, vol. III, Ref. 26 chapters. (Vol. vol. III). Calcutta. Dt. ca CE 16th century: Gurumandala Granthamalaya.

Kumar, K. (2018). *Conservation Management Plan for Kardameshwar Mahadev Temple Precinct, Varanasi;* Unpublished Master of Architecture (Conservation) dissertation. New Delhi: School of Planning and Architecture (SPA–D).

Kumar, M. N. (2010). Land use Patterns and Urbanization in the Holy City of Varanasi India: A Scenario." *Environmental Monitoring and Assessment. Springer Science and Business Media* B.V., pp. 417–422.

Kutlutürk, C. (2013). Significance of Varanasi in terms of Indian religions. *IOSR Journal of Humanities And Social Science, 10*(2), 36–40.
Lannoy, R. (2002). Benares, A World Within a World. In R. Lannoy, Benares, *A World Within a World* (pp. 140–180). Varanasi: Indica Books.
Logan, W. S. (2010). Foreword: Heritagescapes & Cultural Landscapes. In R. P. Singh, *Heritagescapes and Cultural Landscapes* (pp. 1–6). Gurgaon: Shubhi Publications.
Malville, J. M. (2004). Pilgrims to Kashi: Self–Organized Patterns in Space and Time. In P. K. Choudhary, *Sociology of Pilgrims* (pp. 163 – 176). Delhi: Kalpaz Publications.
Motichand. (1985). Kashi ka Itihas (Hindi). In *Motichand, Kashi ka Itihas* (Hindi) (pp. 180–210). Varanasi: Vishvavidyalaya Prakashan, 2 edition.
Nanamoli, B. (1980). The Life of the Buddha. *Buddhist Publication Society, Kandy* (Vol. 1 No. 1 (1976): Pali Buddhist Review), pp. 378. doi: https://doi.org/10.1558/bsrv.v1i1.13756.
Prorok, C. V. (2003). Transplanting pilgrimage traditions in the Americas. *The Geographical Review 93*(3), 283–307.
Rana, S. R. (2019). Visioning Cultural Heritage and Planning: Banaras, the Cultural Capital of India. In B. V. Sandmeier, *Routledge Companion of Global Heritage Conservation* (pp. 152–176). Abingdon: Routledge Taylor & Francis.
Saha, A. (2022). The Evolving Heritagescape in India's Hindu Pilgrimage City of Varanasi: Contemporary Relevance and Contestations. *Heritage & Society, 15*(1), pp 32–45.
Saraswati, B. N. (1975). Sacred–Secular Continuum. In B. N. Saraswati, *Kashi: Myth and Reality of a Classical Cultural Tradition* (pp. 41–47). Shimla: Indian Institute of Advanced Study.
Saraswati, B. N. (1985). Kashi pilgrimage: the End of an Endless Journey. In M. Jha, *Dimensions of pilgrimage* (pp. 91–104). New Delhi: Inter–India Publications.
Schechner, R. (1993). Crossing the Water: Pilgrimage, Movement, and, Environmental Scenography of Ramlila of Ramnagar. In Bradley R. Hertel & C. A. Humes, *Living Banaras, Hindu Religion in Cultural Context* (pp. 1–20). New York: State University of New York Press.
Sherring, M. A. (1868). Source of the great wealth of Benares . In M. A. Sherring, Benares. *The Sacred City of the Hindus* (p. 328). London: Trubner & Co.
Shobana K. (2011, June 2024 25). Gupta–period temples found *Remains of four structures at protected site near Varanasi.* Retrieved from The Telegraph online: https://www.telegraphindia.com/india/gupta–period–temples–found–remains–of–four–structures–at–protected–site–near–varanasi/cid/471740.
Singh, R. (1955). Planning And Improvement of Banaras. In R. Singh, Banaras. *A Study in Urban Geography* (p. 159). Banaras: Nand Kishore & Sons.
Singh, R. P. (1988). The Image of Varanasi: Sacrality and Perceptual world. *National Geographical Journal of India,* 1–16.
Singh, R. P. (1993). Banāras (Vārāṇasī): Cosmic Order, Sacred City, Hindu Traditions : Festschrift to Prof. R.L. Singh. In R. L. Singh, *Issue 5 of Cultural traditions of India* (p. 37). Varanasi: Tara Book Agency.
Singh, R. P. (2002). Environmental! sustainability and issue of Religious heritage. In R. P. Singh, *Towards the Pilgrimage Archetype. Pancakroshi Yatra of Banaras* (p. 187). Varanasi: Indica Books.
Singh, R. P. (2002). Towards the Pilgrimage Archetype. Pancakroshi Yatra of Banaras. *Pilgrimage & Cosmology Series: 3.* Varanasi: Indica Books.

Singh, R. P. (2009). Banaras. In R. P. Singh, Banaras, *The Heritage City of India: Geography, History and Bibliography* (pp. 235–255). Varanasi: Indica Books.

Singh, R. P. (2013). Hindu Tradition of Pilgrimage:. In R. P. Singh, *Hindu Tradition of Pilgrimage: Sacred Space and System* (pp. 225–245). New Delhi: Dev Publishers & Distributors.

Singh, R. P. (2017). Varanasi: Heritage Zones and its Designation in UNESCO's World Heritage Properties. *Kashi Journal of Social Sciences 7* (1–2), *Special issue on "Researching Kashi"* (pp. 201–218). Varanasi: Jhunna Foundation.

Singh, R. P. (2018). Kardameshvara (Kashi) temple, a Religious Heritage from India: Sacred Landscape, Architectural Designs and Perspectives. Sthāpatyam, *Indian Journal of Architecture & Allied Sciences,* pp. 181– 204.

Singh, R. P. (2019). Visioning Cultural Heritage and Planning: Banaras, the Cultural Capital of India. In B. V. Sandmeier, *Routledge Companion of Global Heritage Conservation* (pp. 152–176). Abingdon: Routledge Taylor & Francis.

Singh, U. (2009). Cultural Transition. In U. Singh, *A History of Ancient and Early Medieval India* (pp. 182–255). Chennai: Pearson India Education Services Pvt. Ltd.

Smith, M. P. (1984). Urban Structure, Social Theory, and Political Power. In M. P. Smith, *Cities, Transformation: Class, Capital and the State* (pp. 9–23). USA: Sage Publications.

Sukul, K. (2000). Kashi Tatha Varanasi Ki Yatrayen (in Hindi). In K. Sukul, *Varanasi Vaibhav* (pp. 175–227). Patna: Bihar Rashtra Bhasha Parishad.

Sukul, K. N. (1974). *Varanasi Down the Ages*. Varanasi: Bhargava Bhushan Press.

Tripathi, L. (1962–63). The Kardameshvara temple at Kandwa. Bharati, *Bulletin of the College of Indology,* (B.H.U., Varanasi), No. 6 (pt. 1), pp. 137–156.

Vidyarthi, L. S. (1979). Sacred Specialists: Typology. In L. S. Vidyarthi, *The Sacred Complex of Kashi (A Microcosm of Indian Civilization)* (pp. 91–110). New Delhi: Concept Publishing Company.

Zaidman, N. A. (2001). The marketing of sacred goods: consumers and retailers interaction. *Journal of International Consumer Marketing, 13,* 5–27.

CHAPTER 7

HERITAGE MANAGEMENT AND TOURISM DEVELOPMENT STRATEGIES IN TAMIL NADU

I. M. KARTHIKEYAN

Assistant Professor, Department of Tourism and Hospitality Management, Annai Fathima College of Arts and Science, Thirumangalam, Madurai, Tamil Nadu, India

ABSTRACT

The heritage area is a comparatively new conservation strategy for natural and cultural resources (Edwards & Coit, 1996) that involves both host communities and tourists in the tourism-based telling of the place's story. National Heritage Areas (NHAs) must demonstrate that heritage tourism (Fyall, 2000; Aas et al., 2005; Vela, 2009; Logar, 2010) is compatible with an area's economic activity in order to be designated. The purpose of this study is to provide a methodological structure that complements community-based planning procedures in the following ways: 1) decide the practicality of NHA tourism development from a market evaluation approach, and 2) recognize results-oriented heritage tourism performance measures for assessing the effectiveness of NHA tourism management. Based on a random sample of visitors to Mahabalipuram in Chennai, Tamil Nadu, India, this study compares heritage tourism to mass and ecotourism. The study concludes with a proposed Heritage Tourism Planning framework that integrates this work into the larger NHA planning guidelines that are already in place. This research provides new insights into heritage tourism through statistical

Dynamics of Heritage Management and Tourism Development: Trends, Practices, and Approaches.
Surabhi Srivastava & Ruchika Kulshrestha (Eds.)

research methods that can be used to better inform heritage planning and management in NHAs and other heritage areas around the world.

7.1 INTRODUCTION

Heritage tourism – Tourism at historical landmarks is a well-studied field of study, primarily in management and marketing disciplines. Literature hasn't done much on the growth of heritage tourism in outlying areas, so it can be considered a neglected field. In its broadest sense, it is concerned with the growth of tourism in outlying areas, where the attractions frequently include natural features like wildlife, unique landscapes and ecosystems, built heritage, events, and historic sites (Prideaux, 2002). The desire to construct tourist attractions is frequently a response to the deteriorating economic health of peripheral regions. Peripheral regions have experienced significant changes over the past few decades, primarily as a result of rural residents migrating to urban areas in search of employment opportunities and services that are unavailable in the countryside. Depopulation, demographic aging, the loss of commerce and services, heritage deterioration, and other issues progressively affected these territories. Tourism frequently emerges as one of the most promising initiatives for revitalizing these regions' economies. In fact, tourism has emerged as a frequent and strategic topic in numerous public speeches, written documents, and policies, frequently with the goal of ensuring the long-term viability of the economy of outlying regions. Through the identification of factors and strategies that can affect the success of heritage tourism in its development, the purpose of this paper is to examine the extent to which heritage tourism represents a realistic development strategy for Mahabalipuram, a peripheral municipality in Chennai, Tamil Nadu, India.

7.2 THEORETICAL FRAMEWORK

"Heritage" has been used to describe a wide range of phenomena, including cultural, artistic, archaeological, historical, religious, military, natural, and scenic (Prentice, 1993). Based on directives, charters, and international resolutions, Vecco (2010) looked at how the idea of heritage developed in West European nations. He concluded that the term was characterized by expansion and semantic transfer, leading to a broader definition of heritage, which now encompasses both tangible and intangible cultural and natural heritage that has been neglected for a long time and must be protected. The values and relics that have been passed down through generations, whether they are tangible or intangible, typically fall under this category (Ashworth, 1994; Herbert, 1995; Edwards & Coit, 1996). Masser and co. (1994: 31) asserted that: Heritage is more than just something we want to leave to our children and grandchildren; we also want to fully appreciate and experience it. Our current way of life, our legacy from the past, and our legacy to future generations are all examples of heritage. One of the most comprehensive definitions is provided by Boniface & Fowler (1996). They referred to heritage as "the cultural expression of what makes us what we are, our spiritual DNA," citing Germain Greer.

7.3 GEOGRAPHICAL DESCRIPTION OF MAHABALIPURAM'S MUNICIPALITY

The Tamil Pallava dynasty of the 7th to 9th centuries built Mamallapuram. One of the earliest known examples of Dravidian (South Indian) architecture can be found here, with the majority of the structures made entirely of granite. Mamallapuram is also known as Mahabalipuram, and the name is thought to have come from the Pallava king Mahabali Narasimha Varman I. Today, Mamallapuram is primarily a tourist destination and one of Chennai's main attractions. It is only about an hour from the city thanks to the East Coast Road, which has made it easy to get to. Sadly, restaurants, resorts, amusement parks, people, discarded plastic, and frantic traffic now litter the entire strip. There was a time when you could see the Bay of Bengal almost all the way to Mamallapuram; however, there has been so much development that you can only see the ocean as you get closer to Mamallapuram.

7.4 HERITAGE AREA TOURISM

According to Nurse (2008), heritage tourism is fundamentally a social phenomenon that uses intangible and tangible interactions between visitors and an intentional historic, natural, and cultural resource to tell the destination's story. In actuality, heritage tourism is frequently defined as a contemporary economic activity in which these exchanges take place within the larger tourism market, either as a primary interest or through peripheral exposure (McKercher et al., 2002). Heritage areas would probably not be able to fulfill the educational objectives of heritage storytelling or provide the economic development required to support or justify conservation alternatives if they did not have recreation and tourism.

7.4.1 HERITAGE TOURISM MOTIVATION, SATISFACTION, AND LOYALTY

When determining an area's heritage tourism market's size and characteristics, SPUR has proven to be a unique construct. Compared to previous studies that assumed a homogeneous tourism population, these understandings can improve informing NHA feasibility and crash studies. Well-intentioned heritage attractions may fail to support themselves financially and place a strain on public funds if they do not have a thorough understanding of the tourism market (McKercher, 2001). Understanding the scope of the legacy of the travel industry inspirations might be particularly valuable in the beginning phases of NHA arranging while evaluating the probability that a feasible market exists is significant.

It is essential to assess and monitor the quality of the tourism occurrence in order to comprehend how to maintain a heritage tourism strategy over time or measure the efficiency of tourism management actions. According to Noe & Uysal (1997), a useful metric for determining how well a destination meets visitors perceived experiential needs is the construct of visitation satisfaction, which has been the subject of extensive research in the field of tourism literature. According to Yoon & Uysal (2003), the concept of guest loyalty provides a gauge of how committed customers are to an artifact or, in this case, a destination as a product. Loyal visitors are more likely to replicate their consumer behaviors, such as going back to a favorite target or becoming advocates for the destination by referring others to it *via* word of mouth. These constructs provide methods that NHAs can use to assess the muscle of the heritage tourism market and the quality of interactions between destinations and visitors.

7.4.2 STATEMENT OF THE PROBLEM

In spite of the increasing number of NHA designations, there are no systematic methods for objectively assessing the viability of NHA tourism development taking into account the different characteristics of the heritage market. In addition, there are insufficient analytical and results-oriented methods for assessing the effectiveness of visitation management and conservation of heritage resources. Since visitor perspectives are not taken into account in the current feasibility guidelines, very few NHAs have been able to meet their management and financial sustainability objectives. The success of heritage tourism development, on the other hand, can be predicted with a better understanding of the market's sensitivity. This knowledge is especially important for areas with isolated tourism economies that contribute a significant amount to the gross domestic product, like Mahabalipuram, the subject of this study.

7.4.3 METHODOLOGY

This study's methodology was adapted from previous studies on tourism (Oh, 2001; Yoon & Uysal, 2005) and was put to the test to see if the research hypotheses, which stand for common concerns for the planning and management of NHA tourism, were correct. This work does so by providing practical examples of how these methodologies can be used in other contexts and eventually integrated into a framework for planning heritage tourism. A framework of this kind ought to be able to incorporate the opinions of visitors as well as those of local stakeholders to provide a more comprehensive approach to planning and managing heritage tourism, which, in practice, has largely omitted visitor perceptions. The stakeholder study methods are beyond the scope of this study because they are already well-established and frequently used (NHA, 2003).

A participatory mapping process that was used to define the geographic size of the heritage area and the aspects included was carried out prior to the development of the survey instrument that was used in both studies. The first step in the mapping process was a stakeholder scoping process that used a snowballing strategy to find appropriate stakeholders. This was accomplished by creating lists of qualified individuals in a variety of stakeholder areas, including residents, tourism-related businesses, government policymakers, and cultural tradition bearers. A number generated at random was used to select stakeholders from these categorical lists, and they were invited to the meetings. Respondents were asked to write down important island locations that represented the island's heritage based on its natural, cultural, scenic, and historical characteristics during the meetings.

- **Hypothesis:**
  - **H0:** There is no significant difference between groups and trip spending in heritage tourism.
  - **H1:** There is a significant difference between groups and trip spending in heritage tourism.
- **Sampling:** This study adopted a simple random sampling technique. The sample respondents were bifurcated by three such as Heritage Enthusiasts (106), Eco-Focused Tourists (121), and Heritage Apathetic (119).

The K-means cluster analysis's final cluster centers are shown in Table 7.1. The first group was called "Heritage Enthusiasts" because they said that all five factors related to heritage tourism were very important to their vacation. The second group, on the other hand, gave themselves the moniker "Eco-focused" because they valued only the chance to explore natural settings. They also valued natural history and the remaining culturally related variables only slightly. Due to the fact that the purpose of this study was to compare tourists in terms of natural and cultural heritage, the term "eco-tourist" was avoided due to other loaded meanings. The final group was referred to as "heritage apathetics" due to the fact that historical landmarks were of little significance to their trip, that all other variables were somewhat significant, and that none of these things were particularly significant to their experience.

TABLE 7.1 Ultimate Cluster Centers of Total Sample Based on Measures of Consequence of Heritage Attributes

Importance of Opportunities to	Heritage Enthusiasts (n = 106)	Eco-Focused Tourists (n = 121)	Heritage Apathetic (n = 119)
Learn about local culture	3	2	2
Visit historical buildings	3	2	1
Learn about natural history	3	2	2
Learn about local architectural history	3	2	2
Explore a natural setting	3	3	2

Note: 1: Not important; 2: somewhat important; 3: very important; and 4: extremely important.

The variables concerning the study, the characteristics of their journey, travel sort, trip expenditures, and eagerness to return were the other survey items that were used to characterize each segment following the segmentation process.

All demographic variables, including Gender ($F = 11.44$), educational qualification ($F = 4.71$), family income ($F = 35.15$), and whether travel was for individual heritage ($F = 57.61$), showed significant mean differences between groups ($p < 0.01$). Age, on the other hand, was significantly different to a lesser extent than $p < 0.05$ ($F = 4.51$). The null hypothesis of the first hypothesis, that there was no difference, was rejected because each of these differences was found to be significant. All demographics were significantly different between groups of varying heritage orientations, supporting the alternative H1.

Based on Tukey's HSD posthoc comparisons, compared to heritage apathetics ($x = 0.50$), eco-focused tourists ($x = 0.60$) and heritage tourists ($x = 0.60$) were significantly more females than males ($p < 0.01$). Heritage tourists were the same age as other groups. Heritage tourists were surprisingly less educated ($x = 3.45$; $x = 2.45$; $p < 0.05$) and had a lower household income. Heritage apathetics reported the highest income ($x = 3.09$; $p < 0.01$) than the other segments, and $p < 0.05$ for eco-focused and $p < 0.01$ for heritage ($x = 0.11$;). As was to be expected, heritage tourists reported the highest numbers of individuals traveling to investigate their own personal heritage. When evaluating the other groups ($p < 0.01$), trip spending did not statistically differ between groups despite dissimilarities in income levels. As a result, the null hypothesis of H_2 could not be rejected.

As was to be expected, post-trip behavior and approach toward natural and cultural heritage showed significant inter-group differences with large F values (all F values significant at p 0.01). Post-hoc comparisons showed that heritage tourists outperformed other groups in every category and that eco-focused tourists outperformed heritage apathetics ($p < 0.01$).

7.5 FINDINGS

When compared to erstwhile tourists in a single regional tourism people, the findings of this study provided observed evidence of unique characteristics of heritage tourists as a possible niche form of tourism. In addition, this study demonstrated that, in contrast to previous research (Yoon & Uysal, 2005), the general population sample supported the mediating effect of satisfaction among travel motivations and loyalty. Categorical group assignments based on a tourist's heritage orientation (i.e., how important natural, cultural, and built heritage destination attributes were to the tourism experience) moderated the same latent construct intercepts and paths between constructs. Based on these characteristics, cluster analysis produced three populations: eco-focused for those who scored high only on the natural dimensions, heritage apathetic for those who reported low importance on all three heritage dimensions, and heritage enthusiasts for those who scored high on all three dimensions.

First, of the three types of tourists identified by this study, heritage enthusiasts were the most likely to be traveling for personal heritage, had the lowest household income and had the highest percentage of females. The disparity in female-to-male ratios among the Indian population, which accounted for the majority of tourists in this study, may be to blame for the low level of education and income. Heritage tourism trip expenditures were not statistically different from those of the other groups, despite lower income levels. In addition, those who are interested in heritage were the most willing to pay more for services that help preserve cultural and natural heritage. As a result, they were also the most likely group to research the natural and cultural heritage of their destination prior to their trip. These findings suggest that heritage tourists have strong opinions about the concepts of natural and cultural tourism and are more specific in their choices of travel destinations than other groups.

Second, the major constructs of the study, such as push-and-pull motivation, contentment, and loyalty, were measured using composite, reliable measures by exploratory factor analysis. Heritage enthusiasts demonstrated the strongest overall motivations, with their target selection driven more by their pull motivations. This finding indicates that heritage enthusiasts are least likely to substitute destinations based on internal psychological forces and are most likely to choose a destination based on specific attributes. Heritage enthusiasts, on the other hand, were the least pleased with their travel experience of any group. They were, however, the ones who remained the most steadfast and were the ones who were most likely to inform others and return to their destination.

Thirdly, and most significantly, the factor of heritage orientation moderated path estimates within the chain of motivations, satisfaction, and loyalty, despite the fact that this study confirmed Yoon & Uysal (2005) causal factor structure with less power of estimates. Notably, while eco-focused travelers and heritage enthusiasts only showed a weakly marginal mediating effect, heritage apathetics maintained the burly negative mediating effect of pull motivation on satisfaction. This finding suggests that the non-heritage tourism market follows a more traditional relationship between satisfaction and loyalty, whereas the heritage and ecotourism niche groups do not. It is important to take into account how their motivations affect how they perceive the experience as a whole. Except for eco-focused travelers, for whom push motivation was a strong factor but pull motivation was not, all groups had important direct relationships between push and pull motivations. Despite the fact that satisfaction had a marginally weaker impact on loyalty in the heritage enthusiast group, it continued to play a significant role in all groups.

7.6 CONCLUSION

This study set the stage for future research into heritage tourism as a subset of a larger tourism population. Natural and cultural conservation combined with interpretation is a relatively new and novel approach to resource conservation in this niche. The number of heritage areas continues to rise, which means that systematic planning frameworks like restrictions of tolerable amendments and guest experience and resource protection resource inventory and monitoring programs need to incorporate scientifically sound methodological approaches. New research in resource monitoring is required to ensure that an increase in tourism does not have a negative impact on valuable natural and cultural resources, in addition to gaining a better understanding of the heritage tourism market.

It is abundantly clear from the findings of this paper that heritage tourism is a distinct travel phenomenon in comparison to the mass tourism market, as opposed to other niches of tourism that have been extensively studied in the past. At every stage of the forecast and executive process, visitor experiences and resource conditions must be balanced through sound management if heritage areas as a conservation strategy are to be successful and long-lasting. Assumptions about heritage tourists have been put to the test in this study, which also provided the foundation for the necessary demand-side framework to integrate with existing supply-side destination-specific NHA feasibility methods. The ability of heritage destinations in Mahabalipuram, Chennai, and around the world to be reasonably sustainable should be improved by incorporating visitor perceptions into heritage tourism planning. This will allow heritage destinations to successfully conserve natural and cultural heritage resources.

KEYWORDS

- **cultural conservation**
- **cultural heritage resources**
- **heritage tourism market**
- **heritage tourism planning**
- **loyalty**
- **motivations**
- **national heritage area**

REFERENCES

Anderson, T. W., & Amemiya, T. (1998). The asymptotic normal distribution of estimators in factor analysis under general conditions. *Annals of Statistics, 16*(2), 759–771. https://doi.org/10.1214/aos/1176350834.

Andriotis, K., Agiomirgianakis, G., & Mihiotis, A. (2008). Measuring tourist satisfaction: A factor-cluster segmentation approach. *Journal of Vacation Marketing, 14*(3), 221–235.

Ashworth, G. J. (2010). Consuming heritage places: Revisiting established assumptions. *Tourism Recreation Research, 35*(3), 281–290.

Baker, D. A., & Crompton, J. L. (2000). Quality, satisfaction and behavioral intentions. *Annals of Tourism Research, 27*(3), 785–804.

Brodie, R. J., Coviello, N. E., Brookes, R. W., & Little, V. (1997). Towards a paradigm shift in marketing? An examination of current marketing practices. *Journal of Marketing Management, 13*(5), 383–406. https://doi.org/10.1080/0267257X.1997.9964481.

Byrne, B. M. (2013). *Structural Equation Modeling With AMOS: Basic Concepts, Applications, and Programming*. Routledge. Third Edition, ISBN 9780367230715.

Chen, C. -F., & Chen, F. -S. (2010). Experience quality, perceived value, satisfaction and behavioral intentions for heritage tourists. *Tourism Management, 31*(1), 29–35. https://doi.org/10.1016/j.tourman.2009.02.008.

Curran, P. J., West, S. G., & Finch, J. F. (1996). The robustness of test statistics to non-normality and specification error in confirmatory factor analysis. *Psychological Methods, 1*, 16–29.

Hu, L., Bentler, P. M., & Kano, Y. (1992). Can test statistics in covariance structure analysis be trusted? *Psychological Bulletin, 112*, 351–362.

Huang, S., Shen, Y., & Choi, C. (2015). The effects of motivation, satisfaction and perceived value on tourist recommendation. *Conference on Travel and Tourism Research Association International at Portland,* Oregon, pp.1–12.

Hughes, H. L. (1989). Tourism and the arts: A potentially destructive relationship? *Tourism Management, 10*(2), 97–99. https://doi.org/10.1016/0261-5177(89)90050-2.

Hughes, H. L. (1996). Redefining cultural tourism. *Annals of Tourism Research, 23*(3), 707–709. https://doi.org/10.1016/0160-7383(95)00099-2.

Hui, T. K., Wan, D., & Ho, A. (2007). Tourists' satisfaction, recommendation and revisiting Singapore. *Tourism Management, 28*(4), 965–975. https://doi.org/10.1016/j.tourman.2006.08.008.

Kozak, M. (2001). Comparative assessment of tourist satisfaction with destinations across two nationalities. *Tourism Management, 22*(4), 391–401. https://doi.org/10.1016/S0261-5177(00)00064-9.

Kozak, M. (2002). Comparative analysis of tourist motivations by nationality and destinations. *Tourism Management, 23*(3), 221–232.

Kozak, M., & Rimmington, M. (2000). Tourist satisfaction with Mallorca, Spain, as an off-season holiday destination. *Journal of Travel Research, 38*(3), 260–269.

McKercher, B., & du Cros, H. (2003). Testing a cultural tourism typology. *International Journal of Tourism Research, 5*(1), 45–58. https://doi.org/10.1002/jtr.417.

Mingfeng, Z., & Shenzhen, P. R. (2011). Examining the structural relationships of tourist characteristics and destination satisfaction. Paper presented at the 2010 International Conference on E-Business, Management, and Economics, Hong Kong. *29*(4), pp. 624–636.

Padilla, A., & McElroy, J. L. (2007). Cuba and Caribbean tourism after Castro. *Annals of Tourism Research, 34*(3), 649–672.

Park, D.-B., & R., N. (2013). Structural relationships among service quality, satisfaction, and loyalty: The moderating effects of tourists' motivation. Paper presented at the 3rd International Conference on International Trade and Investment Special Conference Theme: 'Non-tariff measures, the new frontier of trade policy?,' Mauritius. *22*(2), 283–298.

Pine, B. J., Peppers, D., & Rogers, M. (1995). Do you want to keep your customers forever? Harvard Business Press. March–April, 103–114.

Silberberg, T. (1995). Cultural tourism and business opportunities for museums and heritage sites. *Tourism Management, 16*(5), 361–365. https://doi.org/10.1016/0261-5177(95)00039-Q.

Stueve, A. M., Cook, S. D., & Drew, D. (2002). The Geotourism study: Phase 1 executive summary, 24.

Um, S., & Crompton, J. L. (1990). Attitude determinants in tourism destination choice. *Annals of Tourism Research, 17*(3), 432–448.

Weaver, D. B., & Lawton, L. J. (2007). Twenty years on: The state of contemporary ecotourism research. *Tourism Management, 28*(5), 1168–1179. https://doi.org/10.1016/j.tourman.2007.03.004.

Yuksel, A., Yuksel, F., & Bilim, Y. (2010). Destination attachment: Effects on customer satisfaction and cognitive, affective and conative loyalty. *Tourism Management, 31*(2), 274–284. https://doi.org/10.1016/j.tourman.2009.03.007.

PART II

BRANDING AND MARKETING OF HERITAGE TOURISM PRODUCTS

CHAPTER 8

WORLD HERITAGE SITES IN THE TOURIST DESTINATIONS KNOWN AS "PUEBLOS MÁGICOS" (MAGIC TOWNS)

MIRIAM EDITH PÉREZ-ROMERO,[1,2]
MARTHA BEATRIZ FLORES-ROMERO,[3] DONAJI JIMÉNEZ-ISLAS,[4] and JOSÉ ÁLVAREZ-GARCÍA[5]

[1]*Business Management Division, TecNM: Higher Technological Institute of Huichapan, Huichapan, Hidalgo, Mexico*

[2]*Doctoral Program in Sustainable Territorial Development (R015), International Doctoral School, University of Extremadura, Badajoz, Spain*

[3]*Faculty of Accounting and Management, Saint Nicholas and Hidalgo Michoacán State University (UMSNH), Morelia, Mexico*

[4]*Renewable Energy Division, TecNM: Higher Technological Institute of Huichapan, Huichapan, Hidalgo, Mexico*

[5]*Departamento de Economía Financiera y Contabilidad, Instituto Universitario de Investigación para el Desarrollo Territorial Sostenible (INTERRA), Universidad de Extremadura, Cáceres, Spain*

ABSTRACT

Mexico's tourist destinations called Magical Towns have an important historical, cultural, and natural heritage that has been preserved and revalued

Dynamics of Heritage Management and Tourism Development: Trends, Practices, and Approaches.
Surabhi Srivastava & Ruchika Kulshrestha (Eds.)

over time. Among the various attractions of these places, there are some sites recognized worldwide by UNESCO as World Heritage Sites, which have been distinguished as such due to their exceptional universal value and with the purpose of preserving them and making them available for the enjoyment of both locals and foreigners. The objective of this work is to identify the potential of heritage tourism found in the Mexican towns that have the Magical Town designation. The research is descriptive, qualitative, and documentary. It has resorted to the use of documentary research techniques and tools. As a result, it was found that out of the 132 Magical Towns, 6 of them have World Heritage Sites. Also, in 2018, these 6 destinations had 4.86% of the economic units of the total Magical Towns, 4.77% of the employed personnel, and 4.64% of the gross census value added of the tourism activity.

8.1 INTRODUCTION

There are places whose cultural or natural importance is so peculiar that they have been designated as World Heritage Sites; UNESCO (United Nations Educational, Scientific and Cultural Organization) established this category to recognize and protect humanity's tangible and intangible assets; historical monuments, archaeological sites, gardens and parks, as well as cities with unique architecture and culture (UNESCO, 2020). This wealth has triggered the development of tourism activity, turning them into outstanding tourist destinations. The tourist typology developed in these destinations is known as "heritage tourism" and represents a new way of relating tourism and culture (Jiménez, 2019). It is also considered a segment of special interest (Herrera, 2017).

In a particular way, Mexico is a touristically diverse country. It has an important cultural and natural heritage that holds great potential for the touristic exploitation of its endogenous resources (González-Domínguez et al., 2018). With the purpose of offering a complementary and diversified tourism offer to the market (Centro de Estudios Sociales y de Opinión Pública, 2017), the Ministry of Tourism launched the Pueblos Mágicos program in 2001 aimed at quaint and unique places in Mexico (SECTUR, 2001) (Flores et al., 2021). Mexico's Magical Towns is a collection of cities, towns, and communities with unparalleled charm, offering a unique vision of Mexican culture through a blend of history, tradition, nature, architecture, handicrafts, and local life, attributes that make them stand out from other destinations. The purpose of this chapter is to describe the potential for heritage tourism found in Mexican towns with the Pueblo Mágico designation. A descriptive, qualitative, and documentary type of research was proposed, using both techniques and tools of documentary research. As a result, a description of the tourism resources of each of the 6 Magical Towns with World Heritage Sites was made:

8.2 WORLD HERITAGE SITES IN MÉXICO

According to UNESCO (2018), the heritage of humanity is not only a relic of the past, but also inspiration and hope for the future, being a source of knowledge, creativity, and solidarity. The 1972 Convention for the Protection of the World Cultural and Natural Heritage is currently approved by 193 nations; of these, 167 have a site on the World Heritage List, which to date includes a total of 1,157 sites, 900 of which are cultural treasures, 218 natural treasures, and 39 mixed treasures; in addition, 55 sites have been identified as being in danger (UNESCO, 2023a). Mexico is one of the 167 nations that adhered to the World Heritage Convention; it has 35 declarations (3.02% of the world total), 27 are cultural properties (77.14% of the total in Mexico), 6 natural properties (17.14% of the total in Mexico), 2 mixed properties (5.71% of the total in Mexico) (UNESCO, 2023b). Table 8.1 lists the 35 declarations in Mexico, identified by name, location, and type of heritage. Islands and protected areas in the Gulf of California have been designated as both natural and endangered properties.

8.3 WORLD HERITAGE IN THE "PUEBLOS MÁGICOS" TOURIST DESTINATIONS

Mexico's Magical Towns offer a wonderful way to discover the natural and cultural beauty of the country while offering the possibility of different types of tourism. Mexico has 132 Magical Towns and 6 of them offer the possibility of living heritage tourism experiences. Table 8.2 presents the World Heritage Sites in the Magical Towns.

Five of the six World Heritage Sites mentioned in Table 8.2 are classified as cultural properties, and the Monarch Butterfly Biosphere Reserve is classified as a natural property. Magical Towns account for 17.14% of Mexico's 35 declared World Heritage Sites, 18.52% of the cultural properties, and 16.67% of the natural properties. Each site is described below:

TABLE 8.1 Mexico's World Heritage Sites

Name	Location	Type of Assets		
		Cultural Property	Natural Property	Mixed Property
Rock Paintings of the Sierra de San Francisco	Mulegé, Baja California Sur	X	–	–
Whale Sanctuary of El Vizcaino	Mulegé, Baja California Sur	–	X	–
Islands and Protected Areas of the Gulf of California	Los Cabos, Baja California Sur	–	X*	–
Historic Fortified Town of Campeche	Campeche, Campeche	X	–	–
Ancient Maya City and Protected Tropical Forests of Calakmul	Calakmul, Campeche	–	–	X
Revillagigedo's Archipiélago	Colima, Colima	–	X	–
Pre-Hispanic City and National Park of Palenque	Palenque, Chiapas	X	–	–
Archaeological Zone of Paquimé	Casas Grandes, Chihuahua	X	–	–
Central University City Campus of the Universidad Nacional Autónoma de México (UNAM)	Coyoacán, Ciudad de México	X	–	–
Historic Centre of Mexico City and Xochimilco	Cuauhtémoc, Ciudad de México	X	–	–
Luis Barragán House and Studio	Miguel Hidalgo, Ciudad de México	X	–	–
Protective town of San Miguel and the Sanctuary of Jesús Nazareno de Atotonilco	San Miguel de Allende, Guanajuato	X	–	–
Camino Real de Tierra Adentro	Guanajuato, Guanajuato	X	–	–
Historic Town of Guanajuato and Adjacent Mines	Guanajuato, Guanajuato	X	–	–
Hospicio Cabañas	Guadalajara, Jalisco	X	–	–
Agave Landscape and Ancient Industrial Facilities of Tequila	Tequila, Jalisco	X	–	–
Aqueduct of Padre Tembleque Hydraulic System	Otumba, Estado de México	X	–	–
Pre-Hispanic City of Teotihuacan	Teotihuacán, Estado de México	X	–	–
Monarch Butterfly Biosphere Reserve	Angangueo, Michoacán	–	X	–

TABLE 8.1 *(Continued)*

Name	Location	Type of Assets		
		Cultural Property	Natural Property	Mixed Property
Historic Centre of Morelia	Morelia, Michoacán	X	–	–
Earliest 16th-Century Monasteries on the Slopes of Popocatepetl	Cuernavaca, Morelos	X	–	–
Archaeological Monuments Zone of Xochicalco	Miacatlán, Morelos	X	–	–
Historic Centre of Oaxaca and Archaeological Site of Monte Albán	Oaxaca de Juárez, Oaxaca	X	–	–
Prehistoric Caves of Yagul and Mitla	Tlacolula de Matamoros, Oaxaca	X	–	–
Historic Centre of Puebla	Puebla, Puebla	X	–	–
Tehuacán-Cuicatlán Valley: originary habitat of Mesoamerica	Tehuacán, Puebla	–	–	X
Franciscan Missions in the Sierra Gorda	Landa de Matamoros, Querétaro	X	–	–
Historic Monuments Zone of Querétaro	Querétaro, Querétaro	X	–	–
Sian Ka'an Reserve	Felipe Carrillo Puerto, Quintana Roo	–	X	–
El Pinacate and Gran Desierto de Altar Biosphere Reserve	San Luis Río Colorado, Sonora	–	X	–
El Tajin, Pre-Hispanic City	Papantla, Veracruz	X	–	–
Historic Monuments Zone of Tlacotalpan	Tlacotalpan, Veracruz	X	–	–
Pre-Hispanic Town of Uxmal	Santa Elena, Yucatán	X	–	–
Pre-Hispanic City of Chichen-Itza	Tinum, Yucatán	X	–	–
Historic Centre of Zacatecas	Zacatecas, Zacatecas	X	–	–

Source: Elaborated by the authors based on UNESCO (2023b).
Note: *Endangered asset.

8.2.1 PRE-HISPANIC CITY AND NATIONAL PARK OF PALENQUE

Palenque Pre-Hispanic City and National Park is made up of two ancient Mayan villages in the Mexican state of Chiapas, in southeastern Mexico. This archaeologically significant area was designated a World Heritage Site by

TABLE 8.2 World Heritage Sites Present in "Pueblo Mágico" Localities

World Heritage Site (WHS)	Year in Which it Was Declared a WHS	Location		Year in Which it Was Named MT
		Magical Town (MT)	Municipality, State	
Pre-Hispanic City and National Park of Palenque	1987	Palenque	Palenque, Chiapas	2015
Archaeological Zone of Paquimé	1998	Casas Grandes	Casas Grandes, Chihuahua	2015
Agave Landscape and Ancient Industrial Facilities of Tequila	2006	Tequila	Tequila, Jalisco	2003
Pre-Hispanic City of Teotihuacan	1987	San Juan Teotihuacán	San Juan Teotihuacán, México	2015
Monarch Butterfly Biosphere Reserve	2008	Mineral de Angangueo	Mineral de Angangueo, Michoacán	2012
El Tajin, Pre-Hispanic City	1992	Papantla	Papantla, Veracruz de Ignacio de la Llave	2012

Source: Elaborated by the authors.

UNESCO in 1987, and the Mexican government named it an Archaeological Monument in 1993, seeking protection under the Federal Law of Archaeological, Artistic, and Historical Monuments and Sites. Furthermore, the archaeological site is safeguarded by the National Institute of Anthropology and History (INAH) and the National Commission for Natural Protected Areas (UNESCO, 2023c).

Palenque National Park is known for its remarkable architectural and sculptural remains. The site has a total area of 1,780 hectares, of which 9 zones and 1,400 buildings have been recognized. Palenque was one of the largest and most developed Mayan cities of the classic era, demonstrating one of humanity's most significant achievements on the American continent; the city has a planned urban layout, with monumental buildings and some of the largest clearings in the entire Mayan area. Numerous residential sections with housing units, funeral sites, ceremonies, and productive enterprises were located around the administrative and civic ceremonial core. Its style is defined by a high level of refinement, lightness, and harmony. Its architecture features vaulted ceiling buildings, interior sanctuaries, and stucco scenes modeled on columns and walls, among other things; vaulted rooms that connect galleries and T-shaped windows. To the above is added the landscape of the place, which together create a city of unique beauty (UNESCO, 2023c).

Palenque Pre-Hispanic City and National Park are situated in the municipality of the same name. Palenque was incorporated into the Magical Towns program in 2015 (SEGOB, 2020). In this destination, tourists can enjoy attractions, festivities, gastronomy, and handicrafts mentioned below (DATATUR, 2023; SEGOB, 2020; SIC, 2023):

8.2.1.1 ATTRACTIONS

Palenque National Park, Palenque downtown, Bonampak and Yaxchitlán, archaeological zones, Palenque site museum "Alberto Ruz L'Huillier," Palenque cultural house, Aluxes Eco Park, Agua Azul waterfalls, Misol-Ha waterfall, Roberto Barrios waterfall, Welib Ja waterfalls, La Asunción waterfall, Santo Domingo church, tourist community El Panchán, Lak Puj Kul Textile Museum, mainly.

8.2.1.2 FESTIVITIES

The patronal feast of Santa Cruz is on May 3 and the feast of Santo Domingo de Guzmán on August 4.

8.2.1.3 GASTRONOMY

Food like tamales chiapanecos, duck in chirmol, sweated fish, and salpicón of deer; among the drinks are the tascalate and the pozol, and for dessert, there are sweets made with honey.

8.2.1.4 HANDICRAFTS

The embroidered garments, ceramics, and amber jewelry stand out. Table 8.3 shows the number of economic units, employed individuals, and Gross Census Value Added (GVCVA) of the tourism activity in Palenque; it also shows the growth rate (GR) between the different census years and the average annual growth rate (AAGR). The GVCVA is the value of production that is added during the work process, by the creative and transformation activity of the occupied personnel, capital, and organization (production factors), exercised on the materials that are consumed in the realization of the economic activity. In this particular case, it corresponds to tourism activity and is expressed in millions of pesos (SECTUR, 2021). It is observed that the AAGR of the economic units is 8.54%, a value very close to the AAGR of the employed personnel, which is 9.18%. However, the GVCVA's AAGR is 59.85%, with a high growth rate from 2013 to 2018.

TABLE 8.3 Economic Data of Palenque

Year	Economic Units	Economic Units' GR	Employed Personnel	Employed Personnel' GR	GVCVA	GVCVA' GR
2003	502	–	2,017	–	112.552	–
2008	753	50%	3,224	60%	223.633	99%
2013	849	13%	3,944	22%	341.600	53%
2018	1,188	40%	4,981	26%	1,190.310	248%
AAGR (2003–2018)	8.54%	–	9.18%	–	59.85%	

Source: Prepared by the authors based on information from INEGI, Economic Censuses.

8.3.2 PAQUIMÉ ARCHAEOLOGICAL ZONE

The Paquimé Archaeological Zone, also known as Casas Grandes, is one of Mexico's most significant archaeological sites. It lies in the state of Chihuahua, in the Gran Chihuahua area. It is one of the main archaeological complexes in northwestern Mexico. It is thought to contain the archaeological remains of 2,000 rooms organized into groupings of living rooms, workplaces, and stores with courtyards. The most common building material is adobe (also known as raw clay). The pre-Hispanic site has been preserved in three axes: 1) house units, 2) plazas, and 3) ceremonial buildings. It is regarded as the largest archaeological zone illustrating the Chihuahuan Desert's peoples and cultures. It flourished between the years 700 and 1475, reaching its pinnacle in the 14th and 15th centuries. Its architecture serves as an illustration of spatial organization. Paquimé was an important link in the commercial and cultural exchanges between the peoples of the southwestern United States, northern Mexico, and Mesoamerica's great civilizations. The area has an extension of 146 hectares (UNESCO, 2023).

On December 2, 1992, a presidential decree established the Paquimé Archaeological Zone. An edict outlining the boundaries and a buffer zone. It was also designated a UNESCO World Heritage Site in 1998. In addition to

INAH, it is protected by the Federal Law of Monuments and Archaeological, Artistic, and Historic Sites. The buffer zone is safeguarded under the urban development plan (UNESCO, 2023d).

The Paquimé Archaeological Zone is situated in the town of Casas Grandes, a place that was incorporated into the Pueblos Mágicos program in 2015 (SEGOB, 2020). In this destination, tourists can enjoy the attractions, festivities, gastronomy, and handicrafts mentioned below (DATATUR, 2023; SEGOB, 2019a; SIC, 2023; LUGTUR, 2023; PDM, 2022a, b):

8.3.2.1 ATTRACTIONS

Paquimé archaeological zone, Casas Grandes River, Casas Grandes Mountain range, Museum of Northern Cultures, Arroyo de los Monos, Casa de Cultura de Nuevo Casas Grandes, Casa de Cultura Sebastián (19th-century house), Cathedral of Nuestra Señora de la Medalla Milagrosa, Chapel of Señor de la Misericordia, Mata Ortiz Pottery Center, Cueva de la Olla, Cueva de la Golondrina, Janos Biosphere Reserve, Piloncillos Valley, San Diego Hacienda, mainly.

8.3.2.2 FESTIVITIES

Regional Wheat Festival at the end of July, Commemoration of the foundation of the city from September 8 to 16, Feast of the Virgin of the Miraculous Medal celebrated in the second half of November, Commemoration of the Battle of Casas Grandes celebrated at the beginning of March, Meeting of OMAWARÍ indigenous peoples, Bi-national Villista Horseback Riding between February and March, Mata Ortiz Pottery Contest in October, Festival in Casas Grandes in the second half of October.

8.3.2.3 GASTRONOMY

You can enjoy Mennonite cheeses, curds, and cottage cheese, either alone or incorporated into dishes such as burritos and beans with cheese. Another dish is the pork tenderloin tostada and sweets prepared with peaches. Among the beverages is the alcoholic drink sotol, which in a peculiar way in Casas Grandes is served with a dose of rattlesnake venom.

8.3.2.4 HANDICRAFTS: DETAILED CLAY POTS INSPIRED BY THE CERAMIC VESSELS FOUND IN THE ARCHAEOLOGICAL ZONE OF PAQUIMÉ

Table 8.4 shows that the AAGR of the economic units of Casas Grandes is 18.75%, the AAGR of the employed personnel is 51.25%, and the AAGR of the GVCVA is 87.12%. It is important to note that although these percentages are high, the specific data for each year should be noted since it is a fairly small number in comparison with the rest of the Magical Towns being analyzed here.

TABLE 8.4 Economic Data of Casas Grandes, Chihuahua

Year	Economic Units	Economic Units' GR	Employed Personnel	Employed Personnel' GR	GVCVA	GVCVA' GR
2003	3	–	5	–	0.709	–
2008	1	–67%	ND	–	ND	–
2013	9	800%	37	–	7.555	–
2018	12	33%	46	24%	10.592	40%
AAGR (2003–2018)		18.75%	–	51.25%	–	87.12%

Abbreviations: ND: Data not available.

Source: Prepared by the authors based on information from INEGI, Economic Censuses.

8.3.3 AGAVE LANDSCAPE AND FORMER INDUSTRIAL FACILITIES

The agave landscape represents the historical and cultural life in Mexico in pre-Hispanic times; it is characterized by the cultivation of agave, a plant that has been used to make tequila, mezcal, and other products for hundreds of years. On their part, the old industrial facilities, which are also part of Mexican culture, include factories, power plants, refineries, warehouses, and many other places that have contributed to the Mexican economy for many decades and in many ways (Sarmiento, 2015). The location spans 35,019 hectares and is situated between the Tequila Volcano's slopes and the Rio Grande's deep valley. The environment is defined by the cultivation of blue agave, which has been used to manufacture tequila since the 16th century, as well as active distilleries, which highlight the growth of international tequila consumption in the 19th and 20th centuries. The agave culture is now considered part of the national identity. This site also bears testament to the Teuchitlan culture, which changed the land between 200 and 900 AD by building terraces for agriculture, houses, temples, ceremonial mounds, and ball games. (UNESCO, 2023e).

There are different federal, state, and municipal legal instruments that provide protection to this heritage; for example: the Federal Law of Artistic, Historic, and Archaeological Monuments and Sites of 1972, the Cultural Heritage Law of the State of Jalisco, the Regulations for the Protection and Improvement of the Urban Image of Tequila, Jalisco (UNESCO, 2023e).

The agave landscape and historic industrial buildings span three municipalities in Jalisco: Tequila, Arenal, and Amatitán. The municipality of Tequila joined the Magical Towns program in 2003 and in addition to the agave landscape and the old industrial facilities, it has the following attractions for tourists (DATATUR, 2023; SEGOB, 2019b; SIC, 2023; Panoramex, 2023):

8.3.3.1 ATTRACTIONS

Hacienda and distillery Jose Cuervo La Rojeña, parish of Santiago Apostle, National Museum of Tequila, the laundries, Quinta Sauza, Tequila Express train or Jose Cuervo Express, Sanctuary of the Holy Cross, chapel El Calvario, waterfall Los Azules, Cuervo World, Los Abuelos Museum, the municipal palace and main square of Tequila, Quinta Sauza, the temple of the Purisima Concepcion, a volcano of Tequila, hacienda, and distillery San Jose del Refugio, distillery Tequilas del Señor.

8.3.3.2 FESTIVITIES

National tequila fair in the months of November–December which coincides with the patron saint festivities; the day of the Holy Cross celebrated on May 3, a cultural week from April 1 to 15, and "los cantaritos" from June 24 to 29.

8.3.3.3 GASTRONOMY

Menudo rojo, tortas ahogadas made from pork, fried beans, goat or beef birria, meat in its own juice, beef broth sewn with slices of quiote and chunks of chayote, red corn pozole with pork, spinazo with purslane, quelite or nopales, cueritos tostadas, corn sopes, enchiladas jalisquillas, tamales de rajas, strawberry or pineapple, pork in green and red sauce, agavero chicken, tostadas raspadas. Desserts: cajeta sweets, jericalla (baked milk flan), homemade fruit ice cream, tequila sorbet. Drinks: hot fruit punch or corn atole with aromas of vanilla, chocolate, or walnut, rice water or Jamaica, and Tequila.

8.3.3.4 HANDICRAFTS

The barrels of oak or oak wood are decorated and engraved with figures alluding to the agave, the amphorae are lined with pigskin and inscribed with typical phrases, and both products can be filled with Tequila. In addition, there are obsidian handicrafts such as knives and pendants. There are also dolls and figures made from corn leaves; belts made of leather or wood frames; silverware, cloth embroidery, barrel figures, and embossed key rings.

Table 8.5. Shows That the AAGR of the Economic Units is 8.93%, the AAGR of the Employed Personnel is 15.60% and the AAGR of the GVCVA is 74.69%

TABLE 8.5 Economic Data of Tequila, Jalisco

Year	Economic Units	Economic Units' GR	Employed Personnel	Employed Personnel' GR	GVCVA	GVCVA' GR
2003	250	–	648	–	25.932	–
2008	406	62%	1137	75%	53.867	108%
2013	528	30%	1352	19%	81.32	51%
2018	607	15%	2265	68%	335.815	313%
AAGR (2003–2018)		8.93%	–	15.60%	–	74.69%

Source: Prepared by the authors based on information from INEGI, Economic Censuses.

8.3.4 PRE-HISPANIC CITY OF TEOTIHUACAN

The Pre-Hispanic City of Teotihuacan testifies to the pre-urban structures of ancient Mexico and is considered to have been one of the largest cities in the Americas between the first and seventh centuries AD. The city is appreciated as a model of urbanization and planning on a large scale. A significant number of buildings were decorated with mural paintings that symbolize the cosmovision and the environment of the time, in addition to the "talud-tablero" as a characteristic of the civil and religious architecture of the city. In its development, they integrated natural elements of the valley, such as the San Juan River, of which they had to alter its course to cross the Calzada de los Muertos, the main reference axis. On the sides of this road are the Pyramids of the Sun and the Moon, as well as the Templo de Quetzalcoatl (UNESCO, 2023f).

Teotihuacan was designated a UNESCO World Heritage Site in 1987. It is under the care of INAH and is legally protected under Mexico's Monuments and Archaeological, Artistic, and Historic Zones Act of 1972. Additionally, there is a 1988 presidential decree protecting an area of 3,381 hectares. There are also development plans and urban image regulations (UNESCO, 2023f).

The Prehispanic City of Teotihuacan is located in the State of Mexico in the municipality of San Juan Teotihuacan. This municipality became a Magical Town in 2015. It is worth mentioning that this designation is shared with San Martin de las Piramides. The attractions of tourist interest that San Juan Teotihuacán has are (DATATUR, 2023; SEGOB, 2019c; SIC, 2023; México Desconocido, 2023a):

8.3.4.1 ATTRACTIONS

Teotihuacán archaeological zone, Sierra de Tlapachique State Park, Maninal Hills, Barranquilla del Águila River, San Juan River, Carlota or Emperor Bridge, Alemán Mansion, San Francisco Mazapa Church, San Juan Bautista Parish and Ex-Convent, Puxtla Chapel, Church of Santa María Maquixco, the Arches of Tenerife, Pulque Route, Animal Kingdom Zoo, Cathedral of the Divine Redeemer, Botanical Garden of Cactaceae, San Juan Market, Temple of the Lady of the Purification, Museum of the Teotihuacan Murals, Manuel Gamio Museum, Cerro Gordo National Park.

8.3.4.2 FESTIVITIES

Celebration of the Lord of "ECCE-HOMO" on May 8, National Fair of the Prickly Pear in August, Sacred Deposit on September 8, "Día de los Muertos" on November 2, Patron Saint Festivities on November 11, and Tlachinolpan on January 1.

8.3.4.3 GASTRONOMY

Barbecue, carnitas, quail, chicken or mutton mixiotes with nopales, quesadillas, tlacoyos, and sopes. Depending on the season, you can also find chinicuiles, maguey worms, quesadillas of grasshoppers, tamales, or tortas de escamoles. Among the beverages are pulque, cactus, prickly pear, and xoconostle liqueurs.

8.3.4.4 HANDICRAFTS

Wood carvings, pyrotechnics, and others based on precious and semiprecious stones such as alpaca, quartz, onyx, silver, clay, obsidian, and jade.

8.3.4.5 DANCES

The Archileos, Moros y Cristianos, and Serranitos. Table 8.6. Shows that for the Town of San Juan Teotihuacán, The AAGR of the Economic Units is 7.67%, The AAGR of the Employed Personnel is 12.62%, and the AAGR of the GVCVA is 64.22%.

TABLE 8.6 Economic Data of San Juan Teotihuacán, Estado de México

Year	Economic Units	Economic Units' GR	Employed Personnel	Employed Personnel' GR	GVCVA	GVCVA' GR
2003	207	–	503	–	26.388	–
2008	235	14%	955	90%	53.654	103%
2013	283	20%	914	–4%	135.573	153%
2018	461	63%	1519	66%	297.546	119%
AAGR (2003–2018)	7.67%	–	12.62%	–	64.22%	

Source: Prepared by the authors based on information from INEGI, Economic Censuses.

8.3.5 MONARCH BUTTERFLY BIOSPHERE RESERVE

The Monarch Butterfly Biosphere Reserve (MBBR) protects key overwintering sites for Monarch Butterflies that annually return from breeding areas such as Canada. The property protects 8 of 14 overwintering colonies that exist in the oyamel forests of central Mexico, representing approximately 70% of the population (UNESCO, 2023). It has a total area of 56,259 hectares, of which 13,551 hectares are core zone and the remaining 42,707 hectares are buffer zone (Vicente, 2020).

The MBBR is one of the most visited natural protected areas in Mexico and is intensively promoted by the Secretary of Tourism of the governments of the State of Michoacán and Mexico; according to data from SEMARNAT & CONANP (2019), the RBMM is visited by more than 100,000 tourists in the November to March season; while outside of that season, tourism activity is paralyzed (Esquivel et al., 2014). It currently has six accesses open to the public that have official permits to provide tourism services: El Rosario (Ocampo municipality, Michoacán), Sierra Chincua (Angangueo, Michoacán), El Capulín (Donato Guerra municipality, State of Mexico), La Mesa (San José de Rincón, State of Mexico) (Brenner, 2006), Macheros (Donato Guerra municipality, State of Mexico), and Senguio (Senguio, Michoacán) (Godínez et al., 2018).

Mineral de Angangueo is one of the municipalities in which the MBBR is located; this place joined the Magical Towns program in 2012. The following are the tourist attractions of this destination (DATATUR, 2023; SEGOB, 2019; SIC, 2023; México Desconocido, 2023; PDM, 2022):

8.3.5.1 ATTRACTIONS

MBBR, El Rosario Sanctuary, Temple of the Immaculate Conception, San Simón Celador Parish, Parker House, San Simón tourist tunnel, Cruz de Hierro viewpoint, Monument to the Miner, Jesús de Nazaretformer hacienda, dam in the Rondanilla ejido.

8.3.5.2 FESTIVITIES

Monarch butterfly festival between the months of February and March, celebration of the Holy Cross on May 3, San Simón patron saint's day on October 28 the Immaculate Conception on December 8, and the Holy Week festival.

8.3.5.3 GASTRONOMY

some typical dishes are moronga, barbacoa, and carnitas. As for fruits, the endrina and capulín, of which liquors are made, in addition to fruits in syrup.

8.3.5.4 HANDICRAFTS

The main image of the handicrafts is the monarch butterfly, regardless of the material used. There are also toys made of wood.

Table 8.7. shows that for Mineral de Angangueo, the AAGR of economic units is 4.63%, the AAGR of employed personnel is 4.43%, and the AAGR of the GVCVA is 2.34%.

TABLE 8.7 Economic Data of Mineral de Angangueo, Michoacán

Year	Economic Units	Economic Units' GR	Employed Personnel	Employed Personnel' GR	GVCVA	GVCVA' GR
2003	50	–	110	–	4.529	–
2008	62	24%	125	14%	3.098	–32%
2013	72	16%	124	–1%	4.319	39%
2018	87	21%	188	52%	6.223	44%
AAGR (2003–2018)		4.63%	–	4.43%	–	2.34%

Source: Prepared by the authors based on information from INEGI, economic censuses.

8.3.6 PRE-HISPANIC CITY OF EL TAJIN

El Tajin is an important site for Mesoamerican archaeology because it is one of the best-preserved examples of the Epiclassic and early Postclassic prehistoric periods, which occurred between the fall of Teotihuacan and the advent of the Aztec kingdom. When the Aztec empire took over the region, El Tajin was abandoned and partially destroyed. The city stands out for its decorated buildings; the reliefs and paintings narrate their rituals and daily life. The site is protected by INAH; it is officially protected by Mexico's Monuments and Archaeological, Artistic, and Historic Zones Law of 1972, and it was declared an Archaeological Monument Zone by presidential decree in 2001, and it has a Site Management Plan. (UNESCO, 2023g). The Prehispanic City of El Tajin is located in Papantla, a municipality that was named a Magical Town in 2012. The following are the tourist attractions of this destination (DATATUR, 2023; SEGOB, 2019e; SIC, 2023; PDM, 2022d).

8.3.6.1 ATTRACTIONS

Church of Our Lady of the Assumption, Volador Monument, Cristo Rey Chapel, Israel C. Tellez Park, Cristo Rey Chapel, Teodoro Cano Museum, Mask Museum, El Tajin Archaeological Zone, Takilsukut Theme Park, Talhpan Ecological Reserve, Xanath-Papantla Ecopark, Insurgentes Socialistas Waterfalls.

8.3.6.2 FESTIVITIES

Corpus Christi festival is celebrated on June 15 during which the dance of the voladores de Papantla is performed, festival Ninín (Day of the Dead) that takes place from October 28 to November 2, "Day of the Lost Child" on December 7, expo Vanilla begins on Saturday before December 10 and concludes the following Sunday which corresponds to the official date of the cutting of vanilla.

8.3.6.3 GASTRONOMY

Guajolote or hen in mole papanteco (sweet mole), mushroom empanadas, corn totopos with salt or sugar, beans in alchuchut, stuffed bocoles, pipian enchiladas, zacahul, izote flower broth, bean tamales and acoyotol. In sweets, you can find vanilla candies, vanilla atole, and vanilla ice cream, as well as pumpkin candies and almond eggs, mainly.

8.3.6.4 *HANDICRAFTS*

Decorative figures, rosaries, and necklaces made from vanilla pods, glazed ceramic pieces, wooden masks, toys, quexquémes, and sarapes with embroidery, typical clothing of the voladores such as baskets, sandals, fans, hats, flutes, and drums.

8.3.6.5 *DANCES*

Dance of the voladores de Papantla. Table 8.8 shows that for the municipality of Papantla, the AAGR of the economic units is 3.22%, the AAGR of the employed personnel is 2.16% and the AAGR of the GVCVA is 13.48%.

TABLE 8.8 Economic Data of Papantla, Veracruz de Ignacio de la Llave

Year	Economic Units	Economic Units' GR	Employed Personnel	Employed Personnel' GR	GVCVA	GVCVA' GR
2003	421	–	1653	–	135.292	–
2008	484	15%	2204	33%	180.309	33%
2013	540	12%	1966	–11%	195.021	8%
2018	638	18%	2225	13%	427.183	119%
AAGR (2003–2018)		3.22%	–	2.16%	–	13.48%

Source: Prepared by the authors based on information from INEGI, Economic Censuses.

8.4 ECONOMIC RELEVANCE OF MAGICAL TOWNS WITH WORLD HERITAGE ATTRACTIONS

The 6 Magical Towns present in this study, as of 2018 registered 2,993 economic units whose activity is characteristic of tourism, said data represents 4.86% of the total number of units in the 132 Magical Towns. It is also worth mentioning that the number of economic units in these Magical Towns has presented an AAGR of 6.80%, *see* Table 8.9.

TABLE 8.9 Economic Units in the Magical Towns with World Heritage Sites

Year	Palenque	Casas Grandes	Tequila	San Juan Teotihuacan	Mineal de Angangueo	Papantla	Total PM with WHS	Total PM	Percentage of Total Economic Units in PM with WHS
2003	502	3	250	207	50	421	1433	29260	4.90%
2008	753	1	406	235	62	484	1941	41269	4.70%
2013	849	9	528	283	72	540	2281	49510	4.61%
2018	1,188	12	607	461	87	638	2993	61532	4.86%
–	–	–	–	–	–	AAGR	6.80%	6.89%	–

Abbreviations: PM: Pueblos Mágicos; and WHS: World heritage site.

Source: Prepared by the authors based on information from INEGI, Economic Censuses.

Figure 8.1 presents the percentage of economic units for each Magical Town in the study, this percentage is considered 100% of the total economic units of the 6 towns. It is observed that Palenque covers 39.69% and the last position is occupied by Casas Grande with 0.40%.

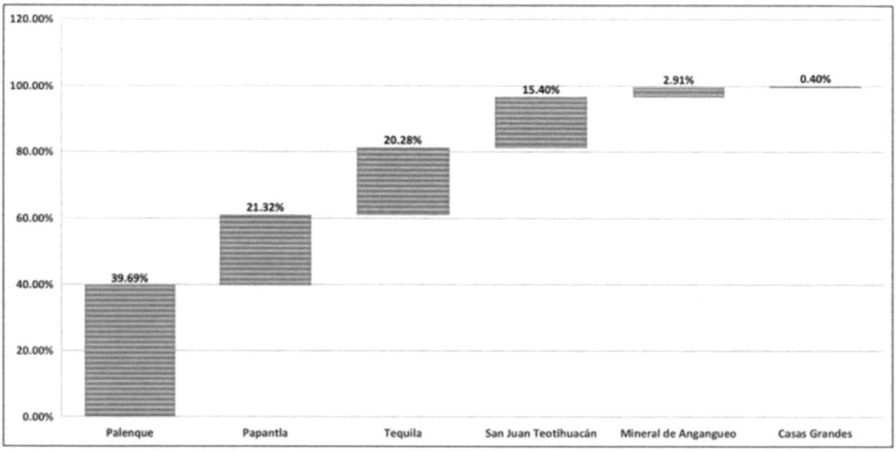

FIGURE 8.1 Economic units as a percentage in the magical towns with world heritage sites.

Source: Prepared by the authors based on information from INEGI, economic censuses.

In terms of employed personnel, the 6 Magical Towns together employ 11,224 people in activities characteristic of tourism, that is, 4.77% of the total in the Magical Towns. The employed personnel show an AAGR of 7.96%, *see* Table 8.10.

TABLE 8.10 Staff Employed in the Magical Towns with World Heritage Sites

Year	Palenque	Casas Grandes	Tequila	San Juan Teotihuacan	Mineal de Angangueo	Papantla	Total, MT with WHS	Total MT	Percentage of Total Employed Personnel in MT with WHS
2003	2,017	5	648	503	110	1653	4936	97,047	5.09%
2008	3,224	ND	1137	955	125	2204	7645	153,632	4.98%
2013	3,944	37	1352	914	124	1966	8337	167,656	4.97%
2018	4,981	46	2265	1519	188	2225	11224	235,518	4.77%
–	–	–	–	–	–	AAGR	7.96%	8.92%	–

Abbreviations: PM: Pueblos Mágicos; and WHS: World heritage site.

Source: Prepared by the authors based on information from INEGI, Economic Censuses.

Figure 8.2 shows that in 2018, Palenque had 44.38% of the employed people among the 6 towns, while the last position is one more time occupied by Casas Grandes, with only 0.41%.

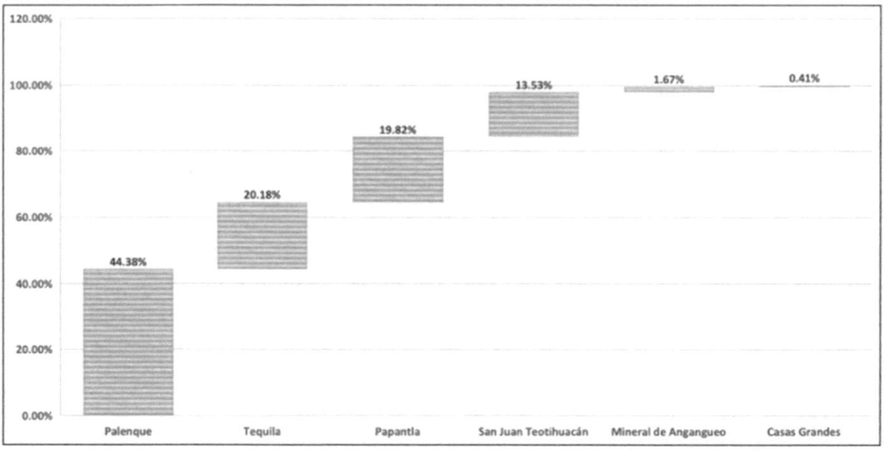

FIGURE 8.2 Personnel employed in percentage in the magical towns with world heritage sites.

Source: Prepared by the authors based on information from INEGI, Economic Censuses.

Regarding the GVCVA, the total of the 6 Pueblos was 2,267.665 million pesos in 2018, a value that represents 4.65% of the GVCVA of the total of Pueblos Mágicos. The AAGR shown by these 6 towns is 40.16%, a value that is almost 2% points above the AAGR of the 132 Magical Towns, see Table 8.11.

TABLE 8.11 GVCVA in the Magical Towns with World Heritage Sites

Year	Palenque	Casas Grandes	Tequila	San Juan Teotihuacan	Mineal de Angangueo	Papantla	Total MT with WHS	Total MT	Percentage of Total GVCVA in MT with WHS
2003	112.552	0.709	25.932	26.388	4.529	135.292	305.402	6,806	4.49%
2008	223.633	–	53.867	53.654	3.098	180.309	514.561	10,835	4.75%
2013	341.6	7.555	81.32	135.573	4.319	195.021	765.388	15,101	5.07%
2018	1,190.31	10.592	335.815	297.546	6.223	427.183	2267.665	48,745	4.65%
–	–	–	–	–	–	AAGR	40.16%	38.51%	–

Abbreviations: PM: Pueblos Mágicos; and WHS: World heritage site.

Source: Prepared by the authors based on information from INEGI, Economic Censuses.

Figure 8.3 shows the GVCVA in percentage for each magical town. Palenque has 52.49% of the VACBT of the 6 towns, while the last position is occupied by Mineral de Angangueo, with only 0.27%.

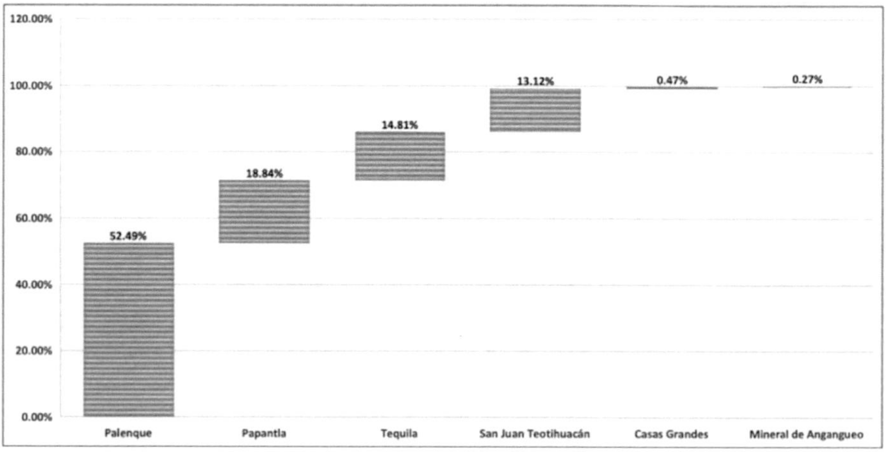

FIGURE 8.3 GVCVA in percentage in the magic towns with world heritage sites.

Source: Prepared by the authors based on information from INEGI, Economic Censuses.

Based on the above analysis, Palenque, whose World Heritage Site is "The Pre-Hispanic City and National Park of Palenque," stands out. It is also worth noting the difference between the towns of Casas Grandes and Mineral de Angangueo.

8.5 CONCLUSION

This chapter describes the potential for heritage tourism that exists in Mexican towns with the Magical Town designation. Out of the 132 Pueblos Mágicos in Mexico, 6 of them have, among their attractions, sites declared by UNESCO as World Heritage Sites. Five of the sites are classified as cultural assets and one site is classified as a natural asset. Mexico's Magical Towns account for 17.14% of its World Heritage Sites. The heritage tourism that characterizes the aforementioned destinations is complemented by attractions, festivities, gastronomy, handicrafts, and in some cases also dances. Recognizing the world heritage sites located in Pueblos Mágicos could facilitate the identification of the segments of tourists to are attracted to these destinations, as well as their expectations and needs.

Of the 6 Magical Towns with World Heritage Sites, Palenque has the highest percentage of economic units (39.69%), the highest percentage of employed personnel (44.38%), and the highest percentage of GVCVA (52.49%). In contrast, Casas Grandes has 0.40% of the economic units and 0.41% of the employed personnel, and Mineral de Angangueo has 0.27% of GVCVA. These data show that the tourist destinations of Casas Grandes and Mineral de Angangueo are the ones that require greater dissemination of their tourist attractions in order to attract more tourists, while at the same time designing strategies to revalue tourism activity in both destinations.

In Mexico, it is characteristic of the design of tourist routes. In that same line, the "Heritage Route of the Magical Towns" could arise, whose main characteristic is the presence of World Heritage Sites. It also has the potential to become a complete geographical adventure, as the tourist who decides to take this route will have to travel through 5 regions of the country: The North-Central Region (Casas Grandes), the Gulf Region (Papantla), the Pacific-South Region (Palenque), the Central Region (San Juan Teotihuacan), and finally, the Central-North and West Region (Tequila & Mineral de Angangueo). In addition to the above, the authenticity that each place offers; the fact that the visitor is in contact with the most authentic aspects of history, culture, or nature is a characteristic that has prevailed in the tourist markets of the last decades.

The study considered a description of the qualities and attributes that could be of interest to tourists. Therefore, future lines of research include the use of quantitative methods such as the valuation of resources through potentiality indexes, a multi-criteria evaluation, or the use of probabilistic tools to develop a regional typification of tourism potential in magical towns.

In addition to analyzing the perspectives of the various stakeholders involved in order to better determine the tourism potential of these destinations, it would also be worthwhile to develop satisfactory and unique experiences in these unique destinations to establish an emotional bond with the tourist and make a difference with respect to others. Finally, it is necessary to understand how the value of the experience and the image of the destination influence the behavioral intentions of visitors, especially in those destinations such as Mineral de Angangueo, which, despite having outstanding attractions, do not have an economic development such as the town of Papantla.

KEYWORDS

- **average annual growth rate**
- **cultural assets**
- **economic development**
- **gross census value added**
- **natural assets**
- **Pueblo Mágico**

REFERENCES

Brenner, L. (2006). Áreas naturales protegidas y ecoturismo: el caso de la Reserva de la Biosfera Mariposa Monarca, México. Relaciones. *Estudios de historia y sociedad, 27*(105), 237–265.

Centro de Estudios Sociales y de Opinión Pública. (2017). Programa de pueblos mágicos: valorización turística de la cultura, documento de trabajo núm. 249. Retrieved from: http://www5.diputados.gob.mx/index.php/camara/Centros-de-Estudio/CESOP/Estudios-e-Investigaciones/Documentos-de-Trabajo/No.-249-Programa-de-Pueblos-Magicos-valorizacion-turistica-de-la-cultura (In Spanish).

DATATUR. (2023). Módulo de información de Pueblos Mágicos. Retrieved from: https://datatur.sectur.gob.mx/PueblosMagicos/pminicio.aspx (In Spanish)

Esquivel, S., Cruz, G., Cadena, C., & Zizumbo, L. (2014). El turismo como instrumento de política ambiental en el Santuario de la Mariposa Monarca El Rosario. *Economía, Sociedad y Territorio, XIV* (44), 141–174. El Colegio Mexiquense, A.C. Toluca, México. (In Spanish)

Flores-Romero, M. B., Pérez-Romero, M. E., Álvarez-García, J., & Del Río-Rama, M. D. L. C. (2021). Bibliometric mapping of research on Magic Towns of Mexico. *Land, 10*(852), 1–24.

Godínez, L., Figueroa, E., Mendoza, M. M., & Pérez-Soto, F. (2018). Análisis del ingreso mensual de las familias en el Rosario, Michoacán. In *Crecimiento y desarrollo sustentable en México*. México: Asociación Mexicana de Investigación Interdisciplinaria A.C. (In Spanish)

González-Domínguez, I., Thomé-Ortiz, H., & Osorio-González, R. (2018). Políticas turísticas y etnoturismo: entre la rururbanización y el desarrollo de capacidades. *PASOS Revista de Turismo y Patrimonio Cultural, 16*(1), 21–36. https://doi.org/10.25145/j.pasos.2018.16.002 (In Spanish).

Herrera, A. (2017). Turismo patrimonial, identidad y desarrollo en el Perú. *Indiana, 34*(1), 199–230. https://doi.org/10.18441/ind.v34i1.199-230 (In Spanish).

Jiménez, C. (2019). El turismo patrimonial: la cultura en venta. In *Hélices y Anclas Para El Desarrollo Local*, 803–812. (In Spanish)

Lugares Turísticos en México-LUGTUR. (2023). Casas Grandes Chihuahua Pueblo Mágico. Retrieved from: https://www.lugaresturisticosenmexico.com/casas-grandes-chihuahua-pueblo-magico/ (In Spanish) (accessed on 18 June 2024).

México Desconocido. (2023a). Artesanías de Obsidiana y Gastronomía. Retrieved from: https://pueblosmagicos.mexicodesconocido.com.mx/estado-de-mexico/san-juan-teotihuacan-y-san-martin-de-las-piramides/atractivo/artesanias-de-obsidiana-y-gastronomia (In Spanish) (accessed on 18 June 2024).

México Desconocido. (2023b). Mineral de Angangueo, Michoacán: atractivos turísticos. Retrieved from: https://www.mexicodesconocido.com.mx/angangueo-mariposas-y-arquitectura-colonial-en-michoacan.html (In Spanish) (accessed on 18 June 2024).

Panoramex. (2023). Fiestas, gastronomía y artesanías. Retrieved from: https://tequila-mexico.com.mx/que-hacer-en-tequila/fiestas-gastronomia-y-artesanias/ (In Spanish) (accessed on 18 June 2024).

Panoramex. (2023). Gastronomía de Tequila. Retrieved from: https://tequila-mexico.com.mx/donde-comer/gastronomia-de-tequila/ (In Spanish) (accessed on 18 June 2024).

Programa Destinos México PDM. (2022b). Artesanías de Chihuahua. Retrieved from: https://programadestinosmexico.com/descubre-mexico/artesanias/artesanias-de-chihuahua.html (In Spanish) (accessed on 18 June 2024).

Programa Destinos México-PDM. (2022a). Gastronomía de Casas Grandes. Retrieved from https://programadestinosmexico.com/descubre-mexico/gastronomia/gastronomia-de-casas-grandes.html (In Spanish) (accessed on 18 June 2024).

Programa Destinos México-PDM. (2022c). Qué comprar en Mineral de Angangueo. Retrieved from: https://programadestinosmexico.com/compras/mineral-de-angangueo-compras.html (In Spanish) (accessed on 18 June 2024).

Programa Destinos México-PDM. (2022d). Pueblos Mágicos: Papantla. Retrieved from: https://programadestinosmexico.com/pueblos-magicos/papantla.html (In Spanish) (accessed on 18 June 2024).

Sarmiento, S. M. (2015). Las antiguas instalaciones industriales en México. Retrieved from: http://www.revista-america.org/revista/index.php/america/article/view/824 (In Spanish) (accessed on 18 June 2024).

SECTUR. (2001). Reglas de operación Pueblos Mágicos. Retrieved from: http://www.sectur.gob.mx/wp-content/uploads/2018/10/Memoria-Documental-Pueblos-M%E2%80%A0gicos-Incorporaci%C2%A2n-y-Permanenecia.pdf (Accessed April 7, 2022). (In Spanish) (accessed on 18 June 2024).

SECTUR. (2021). Primer estudio económico de Pueblos Mágicos. Disponible en DATATUR. (In Spanish)

SEGOB. (2019a). Casas Grandes, Chihuahua. Retrieved from https://www.gob.mx/sectur/articulos/casas-grandes-chihuahua (In Spanish) (accessed on 18 June 2024).

SEGOB. (2019b). Tequila, Jalisco. Retrieved from: https://www.gob.mx/sectur/articulos/tequila-jalisco (In Spanish) (accessed on 18 June 2024).

SEGOB. (2019c). San Juan Teotihuacán y San Martín de las Pirámides, Estado de México. Retrieved from: https://www.gob.mx/sectur/articulos/san-juan-teotihuacan-y-san-martin-de-las-piramides-estado-de-mexico (In Spanish) (accessed on 18 June 2024).

SEGOB. (2019d). Mineral de Angangueo, Michoacán. Retrieved from: https://www.gob.mx/sectur/es/articulos/mineral-de-angangueo-michoacan?idiom=es (In Spanish) (accessed on 18 June 2024).

SEGOB. (2019e). Papantla, Veracruz. Retrieved from https://www.gob.mx/sectur/es/articulos/papantla-veracruz (In Spanish) (accessed on 18 June 2024).

SEGOB. (2020). Palenque, Chiapas. Retrieved from https://www.gob.mx/sectur/articulos/palenque-chiapas (In Spanish) (accessed on 18 June 2024).

Sistema de Información Cultural de México – SIC. (2023). https://sic.gob.mx/resultado.php?busquedaavanzada=Palenque&estado_id=7 (In Spanish) (accessed on 18 June 2024).

UNESCO. (2018). Patrimonio de la Humanidad. Recuperado de https://es.unesco.org/themes/world-heritage/what-world-heritage (In Spanish) (accessed on 18 June 2024).

UNESCO. (2020). Definición de Patrimonio de la Humanidad. Recuperado de https://whc.unesco.org/es/definicion/ (In Spanish) (accessed on 18 June 2024).

UNESCO. (2023a). Patrimonio Mundial. https://es.unesco.org/themes/patrimonio-mundial (In Spanish) (accessed on 18 June 2024).

UNESCO. (2023b). World Heritage List. https://whc.unesco.org/en/list/ (accessed on 18 June 2024).

UNESCO. (2023c). Pre-hispanic city and national park of Palenque. https://whc.unesco.org/en/list/411 (accessed on 18 June 2024).

UNESCO. (2023d). Archaeological Zone of Paquimé, Casas Grandes. https://whc.unesco.org/en/list/560 (accessed on 18 June 2024).

UNESCO. (2023e). Agave Landscape and Ancient Industrial Facilities of Tequila. https://whc.unesco.org/en/list/1209 (accessed on 18 June 2024).

UNESCO. (2023f). Pre-hispanic city of Teotihuacan. https://whc.unesco.org/en/list/414 (accessed on 18 June 2024).

UNESCO. (2023g). El Tajin, Pre-hispanic City. https://whc.unesco.org/en/list/631 (accessed on 18 June 2024).

Vicente, J. C. (2020). Monitoreo de restauración activa y pasiva en un sitio perturbado dentro de la zona núcleo de la Reserva de la Biosfera Mariposa Monarca. Tesis de Maestría en Ciencias Biológicas, Universidad Michoacana de San Nicolás de Hidalgo. (In Spanish).

CHAPTER 9

THE POTENTIAL OF MILITARY TOURISM FOR PEACE FOSTERING AND SUSTAINABLE HERITAGE MANAGEMENT

ISABEL VAZ DE FREITAS[1] and MAKHABBAT RAMAZANOVA[2]

[1]*Professor and Head of the Department of Tourism, Heritage, and Culture, Universidade Portucalense, Portugal*

[2]*Assistant Professor and Coordinator, Tourism and Hospitality Universidade Portucalense, Portugal*

ABSTRACT

The purpose of the preliminary research phase is to provide an overview of the concept of military tourism through a literature review, as the expansion of this type of tourism has had a significant impact on local economies, especially in areas with a strong military history. Military tourism is a concept that explores historical heritage in places of war or conflict. Preserving the identity, cultural values, or integrity of the tangible or intangible heritage of these places becomes crucial. This can be achieved by promoting "sustainable heritage management" with special emphasis on the protection and preservation of heritage for tourism and military purposes. UNWTO (2018) in the tourism agenda for 2030 reinforces the importance of tourism action in the sustainable development goals.

As a second step, it should be explored whether military tourist sites can contribute to peacebuilding. In this way, taking military heritage tourism as an example, tourism can collaborate in an intensive way increasing

Dynamics of Heritage Management and Tourism Development: Trends, Practices, and Approaches.
Surabhi Srivastava & Ruchika Kulshrestha (Eds.)

responsibility for global economic growth (goal 8), encouraging the reduction of inequalities between places and communities (goal 10), developing sustainable cities and communities (goal 11), and particularly can make a difference in peace and justice (goal 16).

Heritage tourism stakeholders can help in reconstruction and regional development after armed conflicts and, at the same time, carry a message of peace. In this way, this sub-sector of heritage and cultural tourism can develop cultural resilience that promotes peace, justice, and mutual understanding, as well as strengthening cultural diversity.

9.1 INTRODUCTION

Tourism and travel are not only associated with leisure, but also with education, knowledge, and sustainable development. In particular, cultural or heritage tourism related to military history is an emerging segment of tourism, which has an important economic, historical, cultural, educational, and strategic value for many countries (Zienkiewicz et al., 2021). Undoubtedly, heritage is a driving force for local development and a driving force for economic, social, and cultural development in cities and rural areas, as an offering that guarantees use and knowledge for future generations (Hassan, Rashed, & Nagar, 2022). In this way, international and national policies created dynamics in the field of heritage management in order to keep alive the history and testimonies of social life through times on Planet Earth. In fact, military tourism is an opportunity to think about regional development and explore heritage conservation by stimulating activities related to leisure, food, hospitality, housing expansion, and the preservation of cultural and natural resources, biodiversity conservation, among others (García-Delgado, Martínez-Puche, & Lois-Gonzaléz, 2020) in places related to military use. It is also an opportunity to reflect on war and all its consequences for social, economic, and cultural life through the meanings and values of heritage. As Díaz-Andreu (2017) argues, the concept of heritage value is linked to heritage protection. In this context, military heritage tourism has a significant task in the communication of the intrinsic values of heritage linked with war and in the sharing of peace messages and social justice.

The current state of research on military tourism demonstrates its broad geographical context and is probably explained by the historical military events that have occurred globally. Most of the research indicates the positive sides of military tourism and its potential for tourism development. Zavarikan (2022) emphasizes the benefits of military tours, mentioning that they have both educational and cultural value for visitors in relation to the conflict situations that have arisen and how they have been solved. Since the last years of the 1980s, the theme of historical military tourism has been worked internationally as a lever for the local and regional development of places of history, heritage, and military culture (Venter, 2017). The first article focuses on tourism as a memorial and empathy for the Second Great War and the concentration camps (Hartmann, 1989). This theme acquires extension around academic studies. Although this concept is not defined with precision, it has been unanimously identified as a subsector of cultural and heritage tourism (Moreno Lobato et al., 2020). Zavarikan (2022) states that this subsector has enormous potential, and its products are attractive due to their connection to highly emotional periods, in particular, moments of war or armed conflict that have a great impact on the tourist or the educational visitor.

9.2 LITERATURE REVIEW

9.2.1 POTENTIAL OF MILITARY TOURISM FOR SUSTAINABLE HERITAGE MANAGEMENT

It is important to reflect on the concept of "sustainable heritage management." Landorf (2009) argues that the participation and empowerment of multiple actors associated with a planning process is one of the principles of sustainable heritage management. The UNESCO Agenda 2030 focuses attention on the integration of culture into development strategies, arguing that this is a way to reduce poverty, increase employment, strengthen the educational process and the acquisition of competencies, promote social justice, and ensure environmental sustainability. Integrating the cultural and heritage needs of protection and conservation into the process of tourism development and public use is the great challenge of sustainable heritage management.

As Rakitovac & Uroševic (2017) conclude, cultural tourism could promote a destination as an "attractive place to live, work, visit, and invest in." Cultural management is also a social challenge for local communities and the local economy. It is an inclusive process involving various actors and stakeholders considering traditional knowledge and habits (Rakitovac & Uroševic, 2017). Tourism is an opportunity for local development, and it is a chance to engage different stakeholders in this development progression using a collaborative and participative approach (García-Delgado, Martínez-Puche, & Lois-González, 2020). It is an opportunity to think sustainably, but some principles must be integrated into the management process. The local community's participation and its sustainable relationship with the natural environment are relevant principles of heritage management (Keitumetse, 2011), as well as social, economic, or cultural policies and practices (Hribar, Bole, & Pipan, 2015). In fact, heritage conservation and preservation are fundamental actions in sustainable place development, but as Luger & Ripp (2020) state, it only makes sense when the local community "creates cultural, social, or economic value from it and benefits accordingly." In this way, the process of heritage management must recognize that local actors should be involved in the research, interpretation, and evaluation process of heritage management in order to protect or find new uses for heritage and culture, developing the potential or creating new opportunities (FakinBajec, 2020).

This connection of heritage or cultural element as a support to tourism points to the need for multidisciplinary teamwork based on history, archaeology, heritage, conservation and restoration, tourism specialists, and, in this case of military heritage in use, the armed forces who are associated with heritage interpretation and management (Figure 9.1).

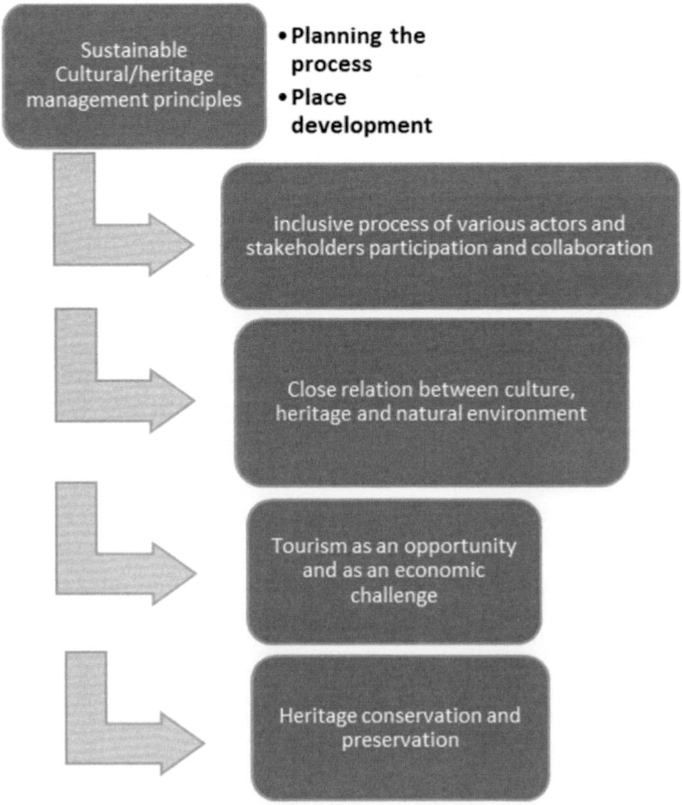

FIGURE 9.1 Sustainable heritage management principles.

The challenge is to balance the preservation of cultural heritage and military identity, while also accommodating visitors, stakeholders, and the local community. In fact, in reviewing the literature, as argued by Zienkiewicz, Podciborski, & Kaźmierczak (2021) and others, cultural and historical events, places, and tangible or intangible heritage related to war or armed activities (Hrusovsky & Noeres, 2011) are key elements for improving and developing heritage conservation and cultural tourism. Two main types of historical-cultural war elements can be distinguished: historical features favorable to battlefields and sites of military-historical significance and living military-historical heritage that is still in use, both offering educational, cultural, and heritage experiences. Particularly, when monuments are still in use and linked to the management of the armed forces, there is a need to discuss how the economics of visiting can cooperate with the sustainable management of heritage, military interests, and use. As such, the military tourist seeks information, experiences, or projects that allow knowing the military historical heritage. In this field of approach, the authors agree with the growing importance of the niche of military, cultural, or heritage tourism, which unites its educational dimensions with historical, ideological, recreational, sports, and adventure, among others.

All material heritage (movable, immovable, archaeological), and immaterial (personalities, battles, war heroes, historical military events, activities, events, legends, and other stories with military content, history of military clothing, values such as identity, reliability, democracy, among others) related to military events are included in this categorization (Rego, Mateus, & Lopes, 2021). On the other hand, this emphasis placed on heritage and culture allows for the existence of a dimension of heritage conservation and rehabilitation (Coelho et al., 2014).

In the areas of military-historical tourism products, there are varied scenarios such as fortresses, castles, museums, events, historical recreations, routes, itineraries, war cemeteries, memorials, exhibitions, military collections, interpreted battle sites, weapons factories, war shelters, bunkers, places related to military leaders or war heroes, among others. In these landscapes, closely linked to war and its settings, the most common themes are military history and heritage. This perspective leads us to military tourism as an animator and experiencer of the historical-military heritage.

Rohrscheidt (2013) associates the literary descriptions of wars, battles, and heroes as one of the possible fields, a perspective that opens up new experiences to the imaginary and emotional values addressed by Chronis (2012), who anchors the imaginary in the experiential field through the narrative that evokes the moral and emotional connection that gives power to tourism.

Still assuming the field of experiences, Halahan et al., (2020) consider that historical and cultural military heritage opens up a perspective in the domains of natural tourism motivated by the search for places of events in natural areas. It is possible to combine historical and cultural interest with the natural scenarios that result in the search for military landscapes, war operations, or itineraries resulting from military activity. This is a more current perspective of territory and sustainability studies that understand military tourism as a tourism sector that acts with an impact on nature conservation. Gatti & Indrigo (2020) state that the preservation and enhancement of the alpine landscape through the use of abandoned World War II paths and cable cars, as support for sports and mountain hikes, discover the perspectives of combining heritage conservation and nature tourism.

9.2.2 POTENTIAL OF MILITARY TOURISM TO PROMOTE PEACEFUL SOCIETIES FOR SOCIAL SUSTAINABLE DEVELOPMENT

Undoubtedly, an emotional approach promotes emotional experiences in cultural and military heritage tourism. Rohrscheidt (2013) refers to the association of memorial culture with military monuments, particularly referring to fortresses, showing the emotional facet conveyed by the duality of the horror of war and the positive value of peace that permeates this type of tourism. There is a strong connection between war, death, and tourism, a landscape of death that inspires the visit (Krokusová et al., 2017). Gatewood (2004) reinforces the emotional component by referring to the fact that tourists are attracted to visit battle sites because they include expressions such as "place of pilgrimage," "hallowed ground" (sacred ground), "national shrine" (sanctuary), among others, transforming these places into memorials and sacred temples, elements of nostalgia, appreciation, values, and emotions. There is an intense relationship between tourism and war in the various heritage, ethnographic, and cultural components, but also the influence of war on tourist movements and tourism in times of war (Gordon, 2018). According to Weaver (2011), after the Second World War, tourism was promoted to mitigate past conflicts and create peace and harmony. It is a new tourism segment that has attracted many visitors in recent years (Zienkiewicz et al., 2021).

Moreno Lobato, Costa, & Hernández-Mogollón (2020) place educational goals and a call for identity at the heart of military cultural and heritage tourism, which is characterized by the creation of innovative products and destinations and culturally sustainable societies through a possible global approach. They refer to the importance of developing certified products and destinations that promote quality products that meet predefined criteria, a challenge to the managers of the historic military monuments and sites.

In particular, military heritage tourism is an opportunity to promote the development of the 16th sustainable development goal that inspires peace, justice, and strong institutions. It is fundamental to think about the protection of heritage in a war field, but it is also fundamental to explore the meanings of cultural heritage as a testimony and evidence of war and conflicts. The meanings of heritage and the power in armed conflicts and occupation mark our days. The studies in that direction appeared in contemporary research (Mozaffari & Barry, 2022). In this process, all stakeholders, including universities, have an important role to develop related to the preservation of identity, authenticity, and the safeguarding of existence in occupied countries (Saif, 2021).

Directly linked with the protection of heritage in war, the Hague Convention (1954) imposes the protection of cultural properties in war places and encourages the States to promote measures in order to protect the cultural heritage, including criminal sanctions for destruction, violations, and theft (UNESCO, 1954). The Hague Convention opened the research on the evident protection of heritage in war and stimulated the emphasis on international practices and methodologies for heritage safeguarding. The current increase in the number of international armed conflicts has heightened international debate about heritage management at conflict sites, conservation, and reconstruction after war (Crooke & Maguire, 2018; Gordon, 2018; McCafferty, 2023; Mohammed & Oumer, 2023; Johnson & Lion, 2022;).

In this approach, "peace through tourism" takes on a new meaning of promoting tolerance, mutual understanding, and the promotion of peaceful societies, strengthening cultural identity, and engaging local communities to prevent violence and conflict (UNWTO, 2018). According to UNWTO (2020), António Guterres' speech reinforces that tourism could promote solidarity and global cooperation, bringing people together, and stimulating peace and sustainability. For Higgins-Desbiolles, Blanchard, & Urbain (2022), tourism is an important means of building inclusive, equitable, and peaceful societies and ensuring the compromise of sustainable development, which opens up new horizons in research and teaching. Erwin & Sturm

(2022) argue that emotive peace tourism could be used as a methodology for reconciliation, to generate tolerance, trust, and empathy based on emotions and peace messages. This emergent model is perfectly in line with the military heritage tourism concept, as it is an emotional subsector of tourism (Gatewood, 2004) based on visitors' emotional experiences (Chronis, 2012). Similarly, Zavarika (2022) believes that tourism is an excellent vehicle for the reconstruction of territories affected by war.

9.3 METHODOLOGY

This study uses bibliometric analysis as a quantitative approach to analyzing literature by examining bibliographic data in a given field of study. Its aim is to identify patterns and trends in research output. It also allows for investigating the concept of military tourism as a method to gain a general understanding of what is known and unknown about this topic, as well as some specific characteristics of the publications.

The bibliometric analysis for this study was conducted during October and November 2022, focusing on academic publications published between 1989 and 2022 in both Scopus and Web of Science as global academic databases. As certain publications were identified in both databases, duplicate entries were removed from the analysis to prevent any potential bias in the results.

The publications were searched using the keywords "military tourism" and military tourism. In the case of using the second keyword (military tourism), we excluded all references that did not address tourism and military perspectives together. We included studies related to other types of tourism associated, directly or indirectly, with military tourism such as war tourism, dark tourism, battle tourism, military cultural tourism, military heritage tourism, military literary tourism, historical-military tourism, or the more recently used term "militourism."

Further, the selected studies underwent analysis based on several criteria of bibliometric analysis such as annual production, country of production, publication type, keywords used by the authors, and keywords in the titles. The aim of this analysis was to identify patterns and trends in the literature related to military tourism and to gain insight into the characteristics and focus of research in this field. In addition, particular emphasis was given to exploring peace and tourism, specifically examining articles that discuss how tourism can contribute to peace-building in areas with a history of military conflict.

9.4 RESULTS AND DISCUSSION

The analysis conducted on studies related to military tourism resulted in the identification of 160 references, being published in journals (130), conference proceedings (16), and books (14) (Figure 9.2).

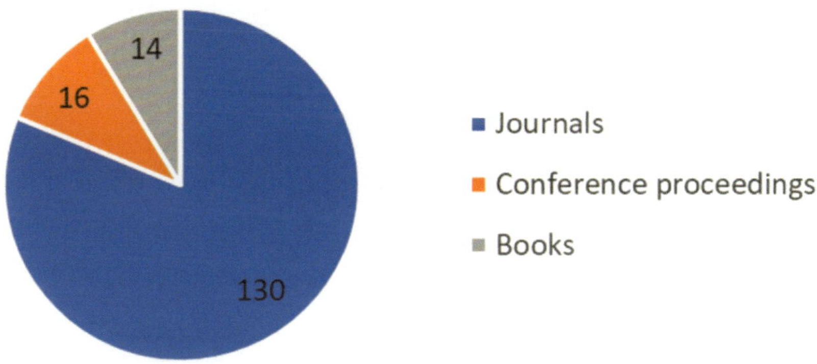

FIGURE 9.2 Publication types.

As for historical analysis, the initial article on this subject was published in 1989, and the number of studies has grown steadily since then, with the highest number of publications being recorded in 2020 (Figure 9.3). This thematic area is relatively new, offering great potential for further research and development.

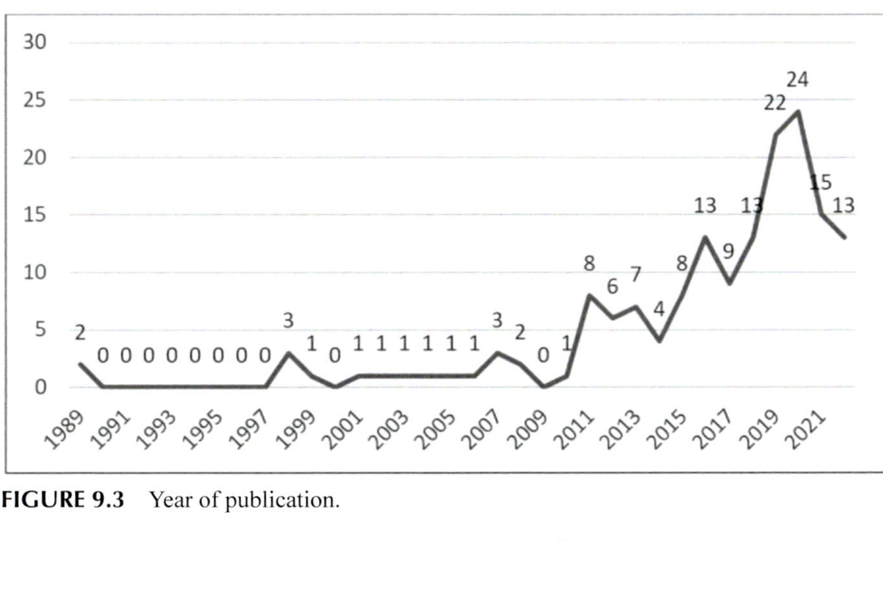

FIGURE 9.3 Year of publication.

In terms of the country of production, the United States has produced the most studies on this topic, followed by the United Kingdom, Poland, Canada, Spain, South Africa, Japan, and several other countries (Figure 9.4).

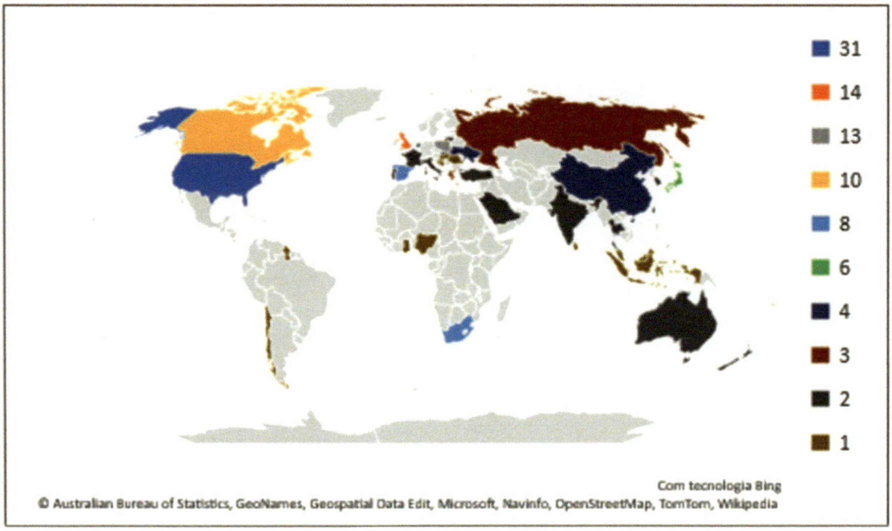

FIGURE 9.4 Country of production.

Analyzing the keywords most used by the authors in the publications. Among the keywords, most used is tourism (38), followed by heritage tourism, a total of 15 times, tourism development a total of 12 times, dark tourism, heritage, and tourist destination cited 8 times, history, terrorism, battle tourism, security, geopolitics, military operations, among others, stand out. Heritage, culture, military, and war were identified as being of significant importance as well (Figure 9.5). The term "sustainability" appears only five times in the selected studies, as does the term "tourism management." This could suggest that sustainability is a relatively recent concern and trend within the field of military tourism. It is likely that the authors have been primarily focused on raising awareness of the topic and educational aspects of military tourism, and not on sustainability aspects. However, the increasing focus on sustainable tourism in general may provide an opportunity for the further exploration of sustainable practices within the context of military tourism, aiming to preserve military heritage sites.

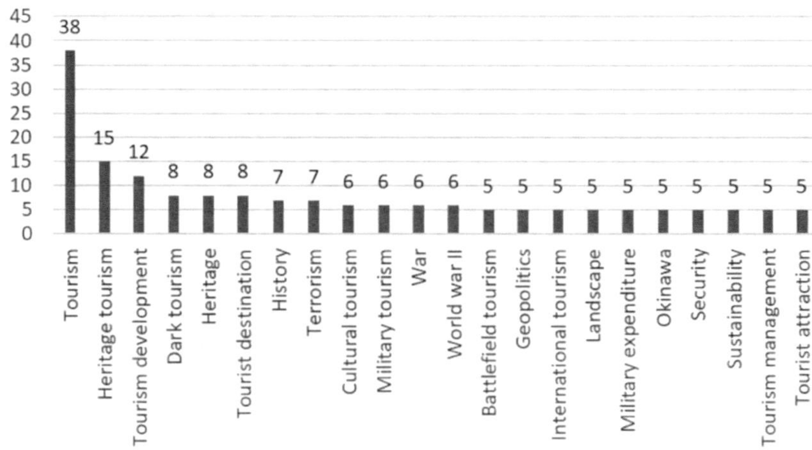

FIGURE 9.5 Keywords mostly used by the authors.

A word cloud created from the titles of the publications confirms the conclusions on the most affected topics of heritage, culture, war, and major battles (Figure 9.6).

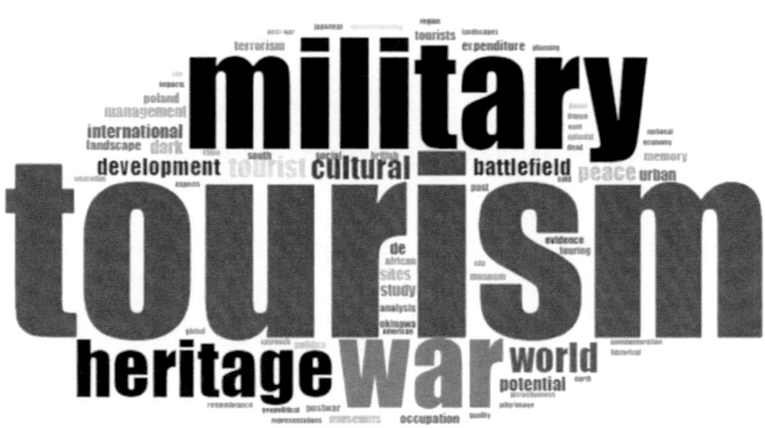

FIGURE 9.6 Word cloud based on the titles of the publications.

Another topic of great importance is the issue of peace promoted by military tourism, which stands out among the themes and is worth mentioning. Other keywords reflect the importance of the landscape, internationalization, and globalization. In this context, Weaver (2011) emphasizes the importance of tourism in the process of understanding war and peace, considering the enormous role of tourism in world peace. In fact, peace associated, in particular, with the capabilities of tourism in bringing people together and its educational messages, is one of the themes to be highlighted and appears prominently in the analysis of the word cloud.

This is how Zavarika (2022) concludes by talking about the importance of the educational and instructive message that trails in wartime landscapes can contain, especially for the younger generation, showing the strong impact of war. Table 9.1 summarizes the themes of historical military heritage (tangible and intangible), its sceneries, and its main dimensions.

TABLE 9.1 Themes of Historical Military Heritage (Tangible and Intangible)

Themes of Historical Military Heritage	Sceneries	Dimensions
Heritage, culture, war, major battles, historical events, places, sites and places of historical-military relevance, personalities, battles, war heroes, historical military events, activities, legends, and other stories with military content, history of military clothing, values (such as identity, reliability, democracy), literary descriptions, narratives.	Fortresses, castles, museums, events, historical recreations, routes, itineraries, war cemeteries, memorials, exhibitions, military collections, interpreted battle sites, weapons factories, war shelters, bunkers, and places related to military leaders or war heroes.	Educational, recreational, heritage conservation and protection, cultural experiences, visit economy, historical experience, ideological, recreational, sports, adventure, emotional experience, and global approach.

9.5 POTENTIAL OF MILITARY TOURISM TO PROMOTE PEACEFUL SOCIETIES

As already mentioned, the article on peace and the role of tourism in the world was selected out of 160 articles. After this selection, a further eight articles were identified and analyzed, and the results of the analysis are presented in Table 9.2. The first article was published in 1999 and deals with the tourism potential for peace dividends. The authors point out that certain types of tourism can bring benefits, but at the same time recognize the difficulties associated with peacebuilding (Butler & Baum, 1999). Interestingly, the rest of the research in this direction has appeared in contemporary studies, being focused on peace education promoted through school tours in the case of Japan's main islands to the site of the tragic battle for Okinawa in 1945 and the site of massive US military bases today (Suzuki, 2016). According to Jung (2019), the prevalence of security tourism on the Korean peninsula could potentially translate into peace tourism today. Asongu et al. (2019) examined how terrorism and peace affected tourist arrivals to 163 countries between 2010 and 2015. The authors emphasize the importance of measuring tourist satisfaction based on their perceptions of peace, security, and terrorism issues.

TABLE 9.2 Peace and Tourism

Country	Title	Year	Journal	Keywords
UK	The Tourism Potential of the Peace Dividend	1999	Journal of Travel Research	–
USA	Bridging Between "Here/Now" And "There/Then": Guiding Japanese Mainland School "Peace Education" Tours in Okinawa	2016	Journal of Cultural Geography	Japan; Okinawa; tourism; education; war; militarism
South Korea	The Construction of the Cold War Landscape and Tourism: The Possibilities and Limitations of Its Peaceful Use	2019	Journal of Asian Sociology	Division border, militarization, strategic village, the Cold War landscape, security tourism, gaze, peace tourism
South Africa	Contemporary Drivers of Global Tourism: Evidence from Terrorism and Peace Factors	2019	Journal of Travel & Tourism Marketing	Terrorism; peace; tourism; generalized method of moments; negative binomial regressions; military expenditure; armed service personnel; security officers and police; drivers and panel data
Canada	Contested Geopolitical Messages for Tourists at the Okinawa Peace Park and Memorials	2019	Japan Forum	Educational dark tourism, geopolitics, Okinawa, Peace Park Museum, storyteller
UK	'Kamikaze' Heritage Tourism in Japan: A Pathway to Peace and Understanding?	2020	Journal of Heritage Tourism	Tourism and peace; Japan; difficult heritage; dissonance; Chiran peace museum; Yūshūkan War Museum
Netherlands	Fortress Tourism: Exploring Dynamics of Tourism, Security, and Peace Around the Virunga Transboundary Conservation Area	2022	Journal of Sustainable Tourism	Peace tourism; conservation; transboundary protected area; militarization
India	Building Peace Through Tourism: the Analysis of an Ongoing Siachen Glacier Dispute Between India and Pakistan	2022	Asian Journal of Comparative Politics	Conflict, India, peace, peace park, Siachen Glacier

More recent studies have focused on the role of international tourism in promoting peace and understanding between peoples and countries (Sharpley, 2020), the role of tourism in promoting peace around the Trans-boundary Protected Areas (Trogisch & Fletcher, 2022), using tourism as a means of reducing current tensions and conflicts by incorporating tourism-peace relations (Wani et al., 2022). It is seen that recent studies have focused on promoting tourism as a means to mitigate past conflicts and promote peace and harmony. The growing interest in this topic can be attributed to the growing trend of tourism development in general. Moreover, with political tensions unfortunately rising, the promotion of military tourism is a potential tool to promote peace, raise awareness among visitors, and preserve military sites as a heritage worthy of attention.

Themes of wars or battles stand out mostly in the contemporary context, where emotionality and experience coexist in tourists' motives (Krokusová et al., 2017). With education and knowledge around military-historical themes as goals, it is also possible to identify the relevant factual importance in promoting peace through the teachings and emotions that spread around war (Weaver, 2011), or still, issues of experience and adventure, particularly in natural spaces as recounted by Halahan et al. (2020).

9.6 CONCLUSION

As a conclusion, the area of military tourism, where conflicts are strong and landscapes are evident, is an opportunity to think about sustainable heritage development, taking into account the multiple themes for investment in innovative visitor products, the variety of landscapes related to heritage and landscapes, and the several dimensions that allow for different interests and expectations of visitors.

Military heritage has a strict link to history and heritage and in relation to cultural or military uses requires specific management strategies. Finally, the relevance of the heritage research context is based on the attempt to identify opportunities for the development of historical and cultural military tourism through increased knowledge, contributing to the memory and identity of national history, including sustainable heritage management practices for visitors.

It is of great importance to identify destinations that can develop military tourism products, as this typology of tourism has enormous potential for study (Venter, 2017). It is important to identify, on a national, regional, or local scale, heritage elements with sufficient power to create military tourism products with a high impact on the development of a sustainable thinking territory. Clear identification of military tourism contexts can open up new tourism products or become a dissipation strategy in cases of tourism massification. The marking of great monuments is an important task, as is their preservation. In this vision of restoration for tourism, there are dangers to heritage, so certification is a way to ensure quality (Moreno Lobato, Costa & Hernández-Mogollón, 2020). In this sense, it is necessary to act with knowledge and quality in the historical and heritage action, but also communicational and educational, since military tourism can be a strong player in the promotion of world peace as pointed out by several authors (Rohrscheidt, 2013). As stated by Sharma et al., (2018), this can be an important point to promote intercultural interaction and understanding between communities and societies globally on the basis of tolerance, inclusiveness, justice, respect, and empathy.

As Higgins-Desbiolles et al., (2022) conclude, promoting tourism as a means for achieving peace not only presents new research opportunities but also holds the potential to promote a sustainable agenda.

KEYWORDS

- **heritage tourism**
- **intercultural interaction**
- **military heritage**
- **tourism management**
- **tourism massification**
- **UNESCO**

REFERENCES

Asongu, S. A., Nnanna, J., Biekpe, N., & Acha-Anyi, P. N. (2019). Contemporary drivers of global tourism: Evidence from terrorism and peace factors. *Journal of Travel & Tourism Marketing, 36*(3), 345–357.

Butler, R. W., & Baum, T. (1999). The tourism potential of the peace dividend. *Journal of Travel Research, 38*(1), 24–29.

Chronis, A. (2012). Between place and story: Gettysburg as tourism imaginary. *Annals of Tourism Research, 39*(4), 1797–1816.

Coelho, J. P., Figueira, L. M., & Costa, C. (2014). O turismo militar e an ativação turística do património: Conceitos, perspetivas e tendências. *Revista Turismo & Desenvolvimento, 1*(21/22), 113–120.

Crooke, E., & Maguire, T. (2018). Negotiating heritage after conflict: Perspectives from Northern Ireland. In E. Crooke & T. Maguire (Eds.), *Heritage after conflict: Northern Ireland,* Routledge, pp. 1–15.

Díaz-Andreu, M. (2017). Heritage values and the public. *Journal of Community Archaeology & Heritage, 4*(1), 2–6. https://doi.org/10.1080/20518196.2016.1228213.

Erwin, J., & Sturm, T. (2022). Living in the wake of rural Irish troubles: Building an institution for sustainable peace through emotive out-of-place tourism. *Journal of Sustainable Tourism, 30*(2–3), 515–532.

Fakin Bajec, J. (2020). An integrated approach to the revitalization, safeguarding and management of cultural heritage: How to establish a durable and active local group of stakeholders. *Annales-Anali za Istrske in Mediteranske Studije-Series Historia et Sociologia, 30*(2), 285–300.

García-Delgado, F. J., Martínez-Puche, A., & Lois-González, R. C. (2020). Heritage, tourism and local development in peripheral rural spaces: Mértola (Baixo Alentejo, Portugal). *Sustainability, 12*(21), 9157.

Gatewood, J. B., & Cameron, C. M. (2004). Battlefield pilgrims at Gettysburg National Military Park. *Ethnology, 43*(3), 193–216.

Gatti, M. P., & Indrigo, A. (2020). The roads, tracks, paths, and ropeways of the First World War: An opportunity to preserve, maintain, and valorize Alpine landscape. *Sustainability, 12*(3), 1157.

Gordon, B. M. (2018). *War Tourism: Second World War France from Defeat and Occupation to the Creation of Heritage.* New York, United States. Cornell University Press. 10.7591/cornell/9781501715877.001.0001.

Halahan, O. O., Kovtoniuk, O. V., Korohoda, N. P., & Braychevskyy, Y. S. (2020). Military landscapes of the Pryvododilni Gorgany as a premise for increasing tourist attractiveness of the area (the case of the Chorna Klyva mountain). *Journal of Geology, Geography and Geoecology, 29*(2), 269–278.

Hartmann, R. (1989). Tourism to the memorial site and museum of the former concentration camp. *Tourism Recreation Research, 14*(1), 41–47.

Hassan, G. F., Rashed, R., & EL Nagar, S. M. (2022). Regenerative urban heritage model: Scoping review of paradigms' progression. *Ain Shams Engineering Journal, 13*(4), 1–8, https://doi.org/10.1016/j.asej.2021.101652.

Higgins-Desbiolles, F., Blanchard, L. A., & Urbain, Y. (2022). Peace through tourism: Critical reflections on the intersections between peace, justice, sustainable development and tourism. *Journal of Sustainable Tourism, 30*(2–3), 335–351.

Hribar, M. Š., Bole, D., & Pipan, P. (2015). Sustainable heritage management: Social, economic and other potentials of culture in local development. *Procedia-Social and Behavioral Sciences, 188*, 103–110.

Hrusovsky, M., & Noeres, K. (2011). Military tourism. In M. Kozak & N. Kozak (Eds.), *The Long Tail of Tourism: Holiday Niches and Their Impact on Mainstream Tourism* (pp. 87–94). Springer. https://doi.org/10.1007/978-3-8349-6231-7_10.

Johnson, J. S., & Lione, B. M. (2022)., Conservation after Conflict: Rebuilding A Heritage Community in Iraq, in C. Smith (Ed.), *The Oxford Handbook of Museum Archaeology* (Oxford Handbooks Online). Oxford, United Kingdom. Oxford University Press.
Jung, K. S. (2019). The construction of the cold war landscape and tourism. *Journal of Asian Sociology, 48*(3), 287–320.
Keitumetse, O. S. (2011). Sustainable development and cultural heritage management in Botswana: Towards sustainable communities. *Sustainable Development, 19*(1), 49–59. https://doi.org/10.1002/sd.419.
Krokusová, J., Miščíková, B., Pasternák, T., & Slivková, S. (2017). The landscapes of death: Relation between war, death and tourism. *Folia Geographica, 59*(1), 36.
Landorf, C. (2009). A framework for sustainable heritage management: A study of UK industrial heritage sites. *International Journal of Heritage Studies, 15*(6), 494–510. https://doi.org/10.1080/13527250903210795.
Luger, K., & Ripp, M. (2020). *Heritage management – Committing to preservation and facilitation: theoretical and practical considerations, In Luger, K., & Ripp, M. (Eds.). World Heritage, Place Making and Sustainable Tourism: Towards Integrative Approaches in Heritage Management. Studien Verlag.*
McCafferty, J. D. (2023). The state heritage institutions of Yemen, Syria and Iraq. In UNESCO, *Cultural Heritage and Conflict in Yemen, Syria and Iraq* (pp. 43–89). Springer International Publishing.
Moreno Lobato, A., Costa, C., & Hernández-Mogollón, J. M. (2020). Value creation of the European Council through the design. *Tourism Planning & Development, 17*(5), 475–495.
Mozaffari, A., & Barry, J. (2022). Heritage and territorial disputes in the Armenia–Azerbaijan conflict: A comparative analysis of the carpet museums of Baku and Shusha. *International Journal of Heritage Studies, 28*(3), 318–340.
Rakitovac, K. A., & Urošević, N. (2017). Valorization of cultural heritage in sustainable tourism. *Management, 12*(3), 199–213.
Rego, C., Mateus, L., & Lopes, E. (2021). The relationship between tourism and the historical-military heritage: Preservation and promotion of the paintings of the military museum of Lisbon. *Journal of Tourism and Heritage Research, 4*(4), 67–77.
Rohrscheidt, A. M. V. (2013). Adaptation of fortress-type sites for cultural tourism. *Ido Movement for Culture. Journal of Martial Arts Anthropology, 13*(2), 38–55.
Saif, Y. (2021). Role of universities in preserving cultural heritage in areas of conflict. *International Journal of Cultural Policy, 27*(7), 866–880. https://doi.org/10.1080/10286632.2020.1857374
Sharma, P., Charak, N. S., & Kumar, R. (2018). Sustainable tourism development and peace: A local community approach. *Journal of Hospitality Application & Research, 13*(1), 36.
Sharpley, R. (2020). 'Kamikaze' heritage tourism in Japan: A pathway to peace and understanding? *Journal of Heritage Tourism, 15*(6), 709–726.
Suzuki, T. (2016). Bridging between "here/now" and "there/then": Guiding Japanese mainland school "peace education" tours in Okinawa. *Journal of Cultural Geography, 33*(1), 100–125.
Trogisch, L., & Fletcher, R. (2022). Fortress tourism: Exploring dynamics of tourism, security and peace around the Virunga transboundary conservation area. *Journal of Sustainable Tourism, 30*(2–3), 352–371.
UNESCO. (1954). Hague convention: Convention for the protection of cultural property in the event of armed conflict. *United Nations Treaty Series, Volume 249.* https://en.unesco.

org/protecting-heritage/convention-and-protocols/1954-convention (accessed on 18 June 2024).

UNWTO. (2018). Tourism in the 2030 agenda. Available at https://www.unwto.org/tourism-in-2030-agenda (accessed on 18 June 2024).

UNWTO. (2020). Tourism can be a platform for overcoming the pandemic. Retrieved from: https://www.unwto.org/news/tourism-can-promote-solidarity-un-secretary-general-antonioguterres (accessed on 18 June 2024).

Venter, D. (2017). Examining military heritage tourism as a niche tourism market in the South African context. *African Journal of Hospitality, Tourism and Leisure, 6*(1), 1–19.

Wani, M. U. D., Dada, Z. A., & Shah, S. A. (2022). Building peace through tourism: The analysis of an ongoing Siachen Glacier dispute between India and Pakistan. *Asian Journal of Comparative Politics, 7*(4), 836–848.

Weaver, A. (2011). Tourism and the military: Pleasure and the war economy. *Annals of Tourism Research, 38*(2), 672–689.

Zavarikan, G. M. (2022). Military tourism as a peculiarity of tourism development in a post-conflict territory. *Journal of Geology, Geography and Geoecology, 31*(1), 186–193.

Zienkiewicz, A., Podciborski, T., & Kaźmierczak, R. (2021). Increased interest in military tourism on selected examples from the area of northern and northeastern Poland. *Communications-Scientific Letters of the University of Zilina, 23*(4), G38–G50.

CHAPTER 10

"I AM A CENTRAL MUSEUM": A SMARTPHONE APPLICATION FOR FILLING THE GAP BETWEEN HISTORY AND VISITORS

SARANG S. DHOTE

Assistant Professor, Department of Chemistry, S.S.E.S. Amravati's Science College, Nagpur, Maharashtra, India

ABSTRACT

"Hello, guests. It was a pleasure meeting you here today. I am Central Museum, Nagpur." The basic idea behind creating the Android app "I am Central Museum" is education via history. It demonstrates an app that offers information about historical monuments and sculptures. The sculptures, historical artifacts, and unique objects themselves give correct information after they scan the QR code or choose the number assigned to each one through this app. The first Central Museum in India to install this kind of app is this one. Marathi, Hindi, and English are the three languages in which this is currently available. Any of these languages may be chosen by users, and the software will function in that language. This app is also functional without an internet connection. This app is copyrighted in the year 2022 in India.

10.1 INTRODUCTION

Museums are the mirror of our culture, heritage, and history. There are many museums established all over the world to represent culture, heritage, and history. Day by day, cultural museums are being added in African and Asian

Dynamics of Heritage Management and Tourism Development: Trends, Practices, and Approaches.
Surabhi Srivastava & Ruchika Kulshrestha (Eds.)

countries for visitors. However, a new challenge for visitors is emerging in museums where many do not have guides to assist visitors. In some cases, a single guide may face a large crowd during weekends, resulting in less information being provided to visitors. For example, recent advancements in virtual technology enable human-system interaction by combining virtual space with a realistic environment and collecting real-time data from people and their surroundings (Zhang, 2018). This technology has attracted significant attention in various industries, such as business (Flavián et al., 2019; Rauschnabel et al., 2019), tourism (Jego et al., 2019), art (Kope et al., 2019), and applications for individuals with special needs (Becker et al., 2018). Historians, curators, and archaeologists are now using virtual technologies to enhance their understanding of historical reconstructions, artworks, and cultural assets (Schofield et al., 2018; Zouboula et al., 2008). However, this VR method is quite expensive, which is why many museums are hesitant to adopt it.

Other forms of technology are needed in these situations. In these circumstances, tagging technologies like Quick Response (QR) codes and Radio Frequency Identification (RFID) have been employed as less expensive options. QR codes are the most widely used of these technologies. This is mostly due to their inexpensive cost, straightforward production method, and low adoption barriers as compared to other comparable technologies. Unlike RFID tags, which can only be read by certain devices, QR codes can be scanned by any smartphone user using a variety of applications and operating systems. The adoption of QR code technologies could revolutionize the visitor experience in settings similar to museums (Mar Pérez-Sanagustín, 2016).

QR codes and VR technologies are two different technologies that serve different purposes. QR Code technology is a 2D barcode that stores information, while VR technology is a computer-generated simulation that allows users to interact with a virtual environment.

QR Code technology has been around for many years and is widely used for various purposes such as product labeling, ticketing, and payment processing. It is a quick and convenient way of sharing information, and its popularity has only increased with the rise of mobile devices. QR codes are easy to generate and scan, and they can be used to store a wide variety of data such as URLs, text, images, and videos. However, QR codes have some limitations. They can only store a limited amount of data, and they require a QR code reader to scan them.

VR technology, on the other hand, is a relatively new technology that is gaining popularity in various industries such as gaming, education, and healthcare. VR technology provides an immersive experience for users, allowing them to interact with a virtual environment as if they were physically present. VR technology requires a VR headset and a powerful computer or gaming console to run. It offers a unique experience that cannot be replicated with any other technology, but it comes with a high cost.

In terms of practicality and versatility, QR Code technology is a more useful and accessible technology than VR. QR codes are simple and inexpensive to create, and they can be used for a wide range of purposes, including advertising, promotions, and information sharing. In contrast, VR technology is still expensive and requires specific hardware to use. It is limited to specific applications such as gaming and simulations, and it is not yet widely accessible to the general public.

However, in terms of immersive experience and entertainment value, VR technology outperforms QR Code technology. VR technology allows users to experience a virtual world in a way that QR Code technology cannot. It offers a unique and interactive experience that can be used for gaming, education, and even therapy.

In conclusion, both QR codes and VR technologies serve different purposes, and it is not fair to compare them directly. QR Code technology is practical and useful for everyday purposes, while VR technology offers an immersive experience that can be used for entertainment and specific applications. Both technologies have their strengths and weaknesses, and it is up to the user to choose the technology that best suits their needs.

The Nagpur Central Museum, also known as the Government Museum, is a treasure trove of artifacts, sculptures, and other archaeological findings. The museum is located in Civil Lines, Nagpur, and is easily accessible from different parts of the city. It is one of the oldest museums in India and is a must-visit for anyone interested in history and culture.

One of the most impressive galleries is the Sculpture Gallery, which houses several stone sculptures from different periods of Indian history. The sculptures are beautifully carved and represent various deities and mythological characters. The gallery also has some rare sculptures from the Mauryan period, which are not found in other museums in India.

Another interesting gallery is the Arms and Armour Gallery, which displays weapons used by different armies in India. The gallery has an impressive collection of swords, shields, bows, and arrows, which were used by soldiers in battles.

Apart from these galleries, the museum also has a Natural History Gallery, which displays different species of animals and birds found in India. The gallery has some rare specimens, including the skeleton of a whale and a crocodile.

One of the most striking aspects of the Nagpur Central Museum is the architecture of the building itself. The museum was built in 1863 and has a colonial-style architecture, with high ceilings, large pillars, and spacious galleries. The building is surrounded by a beautiful garden, which adds to the overall ambiance of the museum.

The museum also has a Numismatic Gallery, which displays coins from different periods of Indian history. The coins are a testament to the evolution of Indian currency and showcase the various empires and rulers that have ruled India over the years.

The museum is spread across two floors and has several galleries that showcase various aspects of Indian history and culture. As soon as you enter the museum, you will be greeted by a massive statue of Lord Buddha, which sets the tone for the rest of the museum. The galleries are arranged in chronological order, so visitors can follow the development of Indian culture and history through the ages.

For decades, visitors to the museum have relied on tour guides or information panels to learn about the exhibits on display. However, with the advent of technology, a new and innovative way of interacting with artifacts has emerged.

In the current work, the author presents their developed smartphone application for the Central Museum, Nagpur where various historical artifacts themselves give information to visitors after scanning the tagged QR code of respective artifacts.

In a recent development, an innovative smartphone application has been developed for the Central Museum, Nagpur. The application makes use of QR code technology to provide visitors with an immersive and interactive experience. With this application, visitors can now scan the QR code tags attached to the exhibits to access detailed information about the artifacts.

10.2 MATERIAL AND METHODS

QR codes have become a popular technology for museums to provide visitors with a more interactive and immersive experience. However, creating and managing these codes can be a daunting task, especially for large museums with numerous artifacts. To simplify this process, many museums have turned to open-source technologies to generate logical QR codes for each artifact. Open-source technologies are software programs that are freely available to the public and can be modified and shared by users. These technologies are often developed and maintained by a community of developers and enthusiasts who contribute to their development. One such open-source technology used by museums to generate QR codes is the ZXing library. This library is a popular QR code generator that can be integrated into software programs to create and manage QR codes. The library is available in several programming languages, including Java, C^{++}, and Python, making it accessible to a wide range of developers. The use of open-source technologies like ZXing makes it easy for museums to generate and manage QR codes for their exhibits. With the help of these tools, museums can create unique QR codes for each artifact, ensuring that visitors have access to relevant and accurate information about the object.

Once the QR codes have been generated, they are printed on regular paper and attached to the appropriate objects. This process is often carried out by museum staff or volunteers who are trained to handle and manage the codes.

The author of the smartphone app "I Am a Central Museum" used the Android Studio platform to develop the app. Android Studio is a popular integrated development environment (IDE) used by developers to create Android applications. The platform provides a range of tools and features that make it easy for developers to create and test apps.

The app developed by the author provides visitors with an immersive and interactive experience. The personalized greeting feature, language choice option, and offline functionality make the app easy to use and accessible to visitors of all backgrounds and languages.

10.3 RESULT AND DISCUSSION

The application provides visitors with information on the history, significance, and cultural context of each artifact. The information is presented in an easy-to-understand manner and is designed to engage visitors of all ages. The application also includes multimedia elements such as images, videos, and audio clips that enhance the visitor's experience.

The software scans each QR code, providing users with fresh and simple ways to learn about historical things. The main museum greets visitors by asking for their names before providing them with a few pieces of information. Language choice is another option for visitors. This smartphone app is available in Marathi, Hindi, and English, and it functions offline and without an internet connection.

One of the key benefits of the application is that it allows visitors to explore the museum at their own pace. Visitors can choose to focus on specific exhibits and learn about them in detail or explore the entire museum at their leisure. The application also provides visitors with the ability to save their favorite exhibits and access them later, making it easier for them to revisit the museum and explore further.

Another advantage of the application is that it provides an eco-friendly solution to the problem of information overload. With the rise of digital technology, museums around the world have faced the challenge of providing visitors with a wealth of information without overwhelming them. The application solves this problem by allowing visitors to access information on demand, without the need for printed materials.

The development of this smartphone application is a testament to the power of technology in enhancing the visitor experience. It provides an innovative and engaging way for visitors to interact with the exhibits and learn about the rich cultural heritage of the region. The application also serves as a model for other museums around the world to follow, as they strive to find new and innovative ways to engage visitors and bring their exhibits to life.

10.4 CONCLUSION

The author can draw the conclusion that using a smartphone-based application can engage every museum visitor from the current work. Also, this cutting-edge technology is useful when there are problems with internet connectivity, language hurdles, or a lack of a guide.

And the development of this smartphone application for the Central Museum, Nagpur is a significant step forward in the realm of museum technology. It is a testament to the power of QR code technology in providing visitors with an immersive and interactive experience. The application not only enhances the visitor experience but also serves as an eco-friendly solution to the problem of information overload. As museums around the world continue to adapt to the changing technological landscape, this application serves as a model for how to engage visitors and bring the exhibits to life.

ACKNOWLEDGMENT

This work is supported by the Archelogy Department Maharashtra India (Nagpur Division).

KEYWORDS

- **android for museum**
- **central museum**
- **cutting-edge technology**
- **museum technology**
- **quick response**
- **radio frequency identification**
- **software**

REFERENCES

App to fill in for guides at city's Ajab Bangla, give info to visitors. (2022). *The Times of India*. https://timesofindia.indiatimes.com/nagpur/app-to-fill-in-for-guides-at-citys-ajab-bangla-give-info-to-visitors/articleshow/91425797.cms (accessed on 18 June 2024).

Becker, G. D. I., Plaza, E., & Besoain, F. (2018). Enhancing spherical photography with videogame visual effects in virtual reality experiences for users with limited mobility. In *Proceedings of the JIISIC 2018-Jornadas Iberoamericanas de Ingenieria de Software e Ingenieria del Conocimiento* (pp. 61–72). IEEE.

Bekele, M. K., Pierdicca, R., Frontoni, E., Malinverni, E. S., & Gain, J. (2018). A survey of augmented, virtual, and mixed reality for cultural heritage. *Journal on Computing and Cultural Heritage, 11*(2), 1–36. https://doi.org/10.1145/3145534.

Bülbül, H. (2017). Müze eğitiminde bir seçenek olarak "second life" müzeleri [using second life museums as an option in education via museum]. *Milli Eğitim Dergisi, 214*, 167–180.

Carrozzino, M., & Bergamasco, M. (2010). Beyond virtual museums: Experiencing immersive virtual reality in real museums. *Journal of Cultural Heritage, 11*(4), 452–458. https://doi.org/10.1016/j.culher.2010.04.001.

Elmqaddem, N. (2019). Augmented reality and virtual reality in education. Myth or reality? *International Journal of Emerging Technologies in Learning, 14*(3), 234–242. https://doi.org/10.3991/ijet.v14i03.9289.

Flavián, C., Ibáñez-Sánchez, S., & Orús, C. (2019). The impact of virtual, augmented and mixed reality technologies on the customer experience. *Journal of Business Research, 100*, 547–560. https://doi.org/10.1016/j.jbusres.2018.10.050.

Gabunilas, L. M., Adlao, E., Burns, K., Chiu, J. S., & Sanchez, S. S. A. (2018). Utilizing portable virtual reality in teaching chemistry. *Science International, 30*(2), 263–266.

Hirose, M. (2005). Virtual reality technology and museum exhibit. In G. Subsol (Ed.), *Virtual storytelling. Using virtual reality technologies for storytelling* (pp. 3–11). Springer.

Jego, L., Gallardo, I., & Besoain, F. (2019). Developing a virtual reality experience with game elements for tourism: Kayak simulator. In *2019 IEEE CHILEAN Conference on Electrical,*

Electronics Engineering, Information and Communication Technologies (CHILECON) (pp. 1–6). IEEE.

Kopeć, W., Wichrowski, M., Kalinowski, K., Jaskulska, A., Skorupska, K., Cnotkowski, D., Tyszka, J., Popieluch, A., Voitenkova, A., Masłyk, R., Gago, P., Krzywicki, M., Kornacka, M., Biele, C., Kobyliński, P., Kowalski, J., Abramczuk, K., Zdrodowska, A., Pochwatko, G., & Marasek, K. (2019). VR with older adults: Participatory design of a virtual ATM training simulation. *IFAC-PapersOnLine, 52*(19), 277–281. https://doi.org/10.1016/j.ifacol.2019.12.110.

Pérez-Sanagustín, M., Parra, D., Verdugo, R., García-Galleguillos, G., & Nussbaum, M. (2016). Using QR codes to increase user engagement in museum-like spaces. *Computers in Human Behavior, 60*, 73–85. https://doi.org/10.1016/j.chb.2016.02.012.

Rauschnabel, P. A., Felix, R., & Hinsch, C. (2019). Augmented reality marketing: How mobile AR-apps can improve brands through inspiration. *Journal of Retailing and Consumer Services, 49*, 43–53. https://doi.org/10.1016/j.jretconser.2019.03.0.

Schofield, G., Beale, G., Beale, N., Fell, M., Hadley, D., Hook, J., Murphy, D., Richards, J., & Thresh, L. (2018). Viking VR: Designing a virtual reality experience for a museum, immersive experiences. In *DIS 2018 – Proceedings of the 2018 Designing Interactive Systems Conference* (pp. 805–816). ACM.

Zhang, Y. (2018). Development of WebGL-based virtual teaching platform for mold design. *International Journal of Emerging Technologies in Learning, 13*(6), 16–28. https://doi.org/10.3991/ijet.v13i06.8581.

Zouboula, N., Fakides, M., & Tsolakidis, C. (2008). Educational uses of virtual reality: Constructing a VR museum. In *International Conference on Interactive Collaborative Learning Proceedings* (pp. 1–13). ICL.

PART III

SOCIAL-CULTURAL ASPECTS OF HERITAGE MANAGEMENT AT TOURIST DESTINATIONS

CHAPTER 11

ETHNIC-BASED HERITAGIZATION AND INTERPRETATION OF PAST CULTURAL LEGACIES IN ETHIOPIA: ITS DISCOURSE AND IMPACTS

TSEGAYE EBABEY and YONAS LEGESSE

Department of Anthropology, Hawassa University, Ethiopia

ABSTRACT

Ethiopia has a living and worth-telling history, as evidenced by enormous cultural and historical legacies. Due to the influence of ethnic politics, the heritagization and interpretation of these legacies have been oriented in an ethnic-based "micro-narrative" perspective since 1991. The purpose of this research is to assess the discourse and subsequent impacts of this conception on historical and cultural legacies, with the goal of reorienting the current trend toward a collective-based heritage approach. By consulting different sources, the study assesses the trends and impacts of this perspective and argues that ethnically oriented cultural heritage has more negative impacts, including degrading the notion of collective heritage, fueling ethnic-prone conflicts that menace the coexistence of Ethiopians, and hindering cultural tourism development through the use of the tremendous cultural endowments. The government's recognition of cultural diversity is merely for political purposes, contrary to community interests. By understanding the negative impacts of this orientation, the heritagization and interpretation of cultural heritage in a collective memory context and the appreciation and utilization of the diverse forms of cultures have to be emphasized to employ

Dynamics of Heritage Management and Tourism Development: Trends, Practices, and Approaches.
Surabhi Srivastava & Ruchika Kulshrestha (Eds.)

heritage for the preservation of the threatened coexistence of Ethiopians and community development practices.

11.1 INTRODUCTION

Ethiopia is a country with a living and worth-telling history (Henze, 2000). This is evidenced by tremendous cultural creativity, including prehistoric human cultural legacies, historic antiquities ranging from precious portable artifacts to architecturally sophisticated and living monuments, and diverse living traditions (Alvarez, 1881; Buxton, 1947; Henze, 2000; Wagner, 2005; Finneran, 2007; Phillipson, 2009; Gosden et al., 2009; McPherron et al., 2010; Joussaume, 2010). Different regimes take different approaches to the processes of heritagization, interpretation, and valorization of these cultural legacies. Currently, this process is proving to be a difficult issue due to the challenge posed by ethno-heritage perception, which contradicts collective heritage conception.

The fracturing of the collective nature of heritage not only makes the heritagization, valorization, and interpretation of cultural legacies a contested area, but it also fuels the cleavage of extreme opinions and ethnic conflicts, as well as degrading Ethiopians' coexistence, sense of belonging, and collective memory in their history. It is used as an instrument of warmongering between different ethnic groups, as has been evidenced in contemporary ethnic and religious tensions and conflicts that employ modern war weapons that the country's resources couldn't shoulder (Markakis, 1989). The discourse of politics towards cultural heritage in general and the trend of ethnic-based heritage interpretation and its impact, in particular, is an unexplored area. This chapter aims to assess the trend, discourse, and subsequent impacts of the anachronistic ethnic-based conception on Ethiopian history and its cultural legacies. Furthermore, it earnestly seeks the prospects of reorienting the heritagization and interpretation of the Ethiopian cultural past as a collective heritage to enable its positive impact either to enhance the fragile socio-economic development or salvage and preserve Ethiopians' coexistence and solidarity for now and the future. To this end, sources from different publications, including articles, books, theses, and reliable internet websites, have been consulted.

11.2 CULTURAL HERITAGE: A CONTEXTUAL REVIEW

The term cultural heritage is a complex concept. This short review articulates some of the aspects related to the scope of this study. Culture, a central aspect of the term, is a dynamic means of coping mechanism that humans use to survive in a harsh environment. It is created, learned, and shared by a person as a member of a community (Cohen, 2001; Kottak, 2015). It exhibits diversity as humans live in different environmental contexts. Its diversity is a source of shared history and pride, and it contributes to the development of a sense of unity and belongingness. It is perceived as humans' common heritage that has benefits for all (UNESCO, 2015; Howard, 2021). The cultural footprint of human history in its inheritance form is defined as cultural heritage, which is everything with irreplaceable value that someone wants to collect, conserve, and pass on to future generations. Heritage implies a process by which the past is used. It is perceived as a present-centered resource with multiple producers and uses managed for present needs. It is interwoven within the life and power dynamics of society and identity construction (Ashworth, Graham, & Tunbridge, 2007; Harvey, 2008). Cultural heritage connotes a concept of both tangible and intangible aspects of the cultural environment, encompassing landscapes, built urban environments, historic places, ancient monuments, archaeological sites, collections, disappeared and continuing cultural practices and arts, ideology, experiences, and knowledge (ICOMOS, 1999; Howard, 2003; UNESCO, 2003; Timothy & Nyaupane, 2009; Chhabra, 2010).

Cultural heritage varies in form and style from place to place as the environment where humans live and the behaviors that they acquire vary. It is, however, defined as a priceless common (public) heritage of humanity and a component of national cultural heritage as a legacy of the past (Merryman, 1986). By combining tangible and intangible assets, cultural heritage is better understood holistically (Albert, Bernecker, & Rudolff, 2013). In the earliest approach to heritage, mainly in the realm of UNESCO, emphasis was given to its tangible aspect. However, following the criticisms of this orientation, intangible heritage has been emphasized mainly since the 2003 Convention of UNESCO (Aikawa-Faure, 2009). In cultural heritage studies, there are three basic interrelated concepts to be pointed out here: heritagization, valorization, and interpretation. Heritagization implies the process of creating communities' feelings of worthiness toward their cultural heritage. It seeks a process of transforming material cultures and practices into a heritage (Sjöholm, 2016; Nilsson, 2018).

Valorization also involves a process of identifying worthy values to enable practices within the concept of heritage (Avrami, Mason & de la Torre, 2000). In any aspect of heritagization, conservation, and tourism practices, valorizing cultural heritage is significant. The idea of conservation and use of past legacies relies on people's attribution of values (esthetic, religious, social, historical, symbolical, economic, and educational values) to heritage. Political entities may attach a political value to using heritage to legitimize their political power or to delegitimize the power of others (Tunbridge & Ashworth, 1996; Mason, 2002; Worthing & Bond, 2008). From the standpoint of tourism economics, cultural heritage is valued in order to generate income for current needs and to serve the needs of future generations (Mason, 2002) by revitalizing the past in order to build a common shared future (Harrison et al., 2020). Cultural heritage, in particular, is a ubiquitous tourism resource being bargained for the improvement of developing-country communities' livelihoods. The growth of heritage tourism as a principal form of the tourism industry and as an important potential tool for poverty alleviation and community development is mainly through reliance on cultural elements of the past (Jamieson, 1998; Srivastava, 2019).

Heritage interpretation is an art form of educational activity that seeks explanation, elucidation, and explication of the meaning of heritage with the aim of understanding, appreciating, and protecting cultural heritage (Nowacki, 2012). It reveals the role of heritage in the development and construction of identities (Smith, 2004). It shapes the image of the public's vision of the past. The way that heritage is interpreted affects the valorization and usage of cultural heritage. It helps to enhance sustainable community planning and heritage protection, public appreciation, understanding of the wider context and authenticity of cultural heritage, and stakeholders' engagement in heritage affairs (ICOMOS, 2008; Silberman, 2013). It is a public discourse that deals with collective identities and social norms represented through heritage ranging from local to international views (Howard, 2003; Nowacki, 2021).

By involving local and larger public cultural contexts, interpretation provides multiple contextual meanings of heritage. This implies interpreting culture as a layered cake that resembles the local, national, and global contexts but seeks to outshine them in their wider context with a shared value of the past. The interpretation that principally relies on local (ethnic) context subjects' cultural heritage to lose its wider context and collectiveness (During, 2011). Rapid political changes, which subject past narrations and legacies to being misinterpreted to achieve claims of different interests or propagate one's political ideology (Howard, 2003; Hassner, 2009; Chhabra, 2010), heritage interpretation is likely subjective. Cultural heritage can be politically misused and abused when history is purposefully manipulated to address different political interests (Galaty & Watkinson, 2004; Hassner,

2009). In the context of ethno-politics, this approach accompanies the racial categorization of people with different cultural backgrounds. A good example of how cultural heritage can be misinterpreted and misused in this context is the case of Nazi Germany, which did employ heritage interpretation for political purposes (Trigger, 1996).

11.3 THE "META-NARRATIVE" APPROACH TO CULTURAL HERITAGE UNDER THE IMPERIAL AND DERG REGIMES OF ETHIOPIA

Human history involves the creation, heritagization, and preservation of cultures. What we inherit today is humans' creativity from the past. Cultural creativity in Ethiopia extends over millions of years, from the beginning of the early Stone Age to the recent past. Heritagization, including ownership and restoration activities, grew mainly after Imam Ahmad's war (1529–1543), which devastated much of the Christian heritage. The restoration made at Aksum Tsion (which was destroyed during the war) by Emperor Fasiledes (r. 1632–1667) can be an example of the concept of heritagization in the early periods (Munro-Hay, 1991). The emperor's acquisition of "Kurate Re'su," a Gondarine painting stolen during Robert Napier's expedition against Emperor Tewodros II in 1868, as a symbol of the monarchy during the Gonderine Period (Pankhurst, 1982), and the documentation of the royal cultural treasures as edifying for future generations by King Sahele Sellasie (1813–47) of Shewa, imply the concept of heritagization (Nair, 2016, pp. 102, 103). Since the reign of Emperor Tewodros II, three important developments have facilitated the intensive heritagization process of cultural legacies. The first was related to the state (re)building process of "Greater Ethiopia" through the construction of a "meta-narrative," a collective heritage conception (Finneran, 2013). Antiquities were collected, documented, and heritagized by emperors as royal treasures and symbols of the monarchy. The collection of Tewodros II at Mekdela is a good example of the state-based heritagization process in the early modern era (Pankhurst, 1973).

During the reign of Emperor Menelik II, the foundation of Addis Ababa as the capital of the empire enabled it to become a center of culture and education that played a major role in the development of the "meta-narrative" (Marzagora, 2017). The archaeological expeditions made possible by the emperor's willingness were part of this development (Ashenafi, 2018). The process of heritagization based on a well-defined narrative was practiced during the reign of Emperor Haile Selassie I (Marzagora, 2017). After the Italo-Ethiopian War (1935–41), the regime was effective in heritagizing cultural legacies as signs of "Greater Ethiopia." Ethiopia became a strategic partner of UNESCO in the 1960s and served as a voice for African countries abroad. The regime was concerned with promoting Ethiopia as an "African State" and placing it within a brilliant global image through the creation of a "heritage meta-narrative" that played a major role in the inscription of many of the world's heritages (Finneran, 2013; Huber, 2017, 2021).

The second development was associated with the Victory of Adwa and its post-deployments. The victory not only guaranteed Ethiopian sovereignty within colonized Africa but also allowed it to establish diplomatic relations with the colonial powers at the time. It also promoted the country worldwide as a symbol of freedom and the center of early African civilization through the movement of Pan-Africanism. In this regard, the historic monuments were considered representative of African civilization. Pan-Africanism gradually gave birth to the Organization of African Unity (OAU), which is now known as the African Union (AU). Haile Selassie, I strategically adjusted infrastructures to make Addis Ababa the headquarters of the AU and other world organizations and a center of diplomacy. This was an important development for the culmination of "Greater Ethiopia" (Marcus, 1966; Huber, 2017). The third development was associated with the progress in foreign-based archaeological and historical investigations that left records helpful in the inscription process of the world heritage sites of Ethiopia. The developments overall have contributed to Ethiopia's heritagization "boom years" of the 1960s and 1970s. Of the first twelve nominated world heritage sites, two of them (the Rock Churches of Lalibela and the Semien Mountains National Park) were selected from Ethiopia. In the first three UNESCO sessions, seven sites were inscribed. This was a big achievement of the period that needs recognition (Deacon, 2014; Nair, 2016; Huber, 2021). However, the heritagization process of this regime had some limitations (see Nair, 2016; Huber, 2021). Firstly, the process of heritagization was a "top-down" approach that excluded major stakeholders (such as the Ethiopian Orthodox Church) and local communities from the protection and ownership of cultural resources found around them. The legislation of the regime did not clearly state the role of major stakeholders in cultural heritage. Ambiguously, all antiquities found in the country were stated to be owned by the state (though this was done to protect the illegal movement of antiquities). This slowed the active participation of local communities by assuming that managing cultural heritage was solely the responsibility of the state.

Secondly, due to the high involvement of foreign experts, the heritagization process was influenced by the Western cultural heritage perspective led by UNESCO, which gave more emphasis to physical monuments and artifacts. This contributed to the exclusion of intangible cultural values from the heritagization process. The concentration made on the tangible "historic route" antiquities in the north was partly associated with the global trend that primarily gave emphasis to the outstanding universal values of monumental and architectural features. This perspective was dominantly led by Western scholars, both at home and abroad. Thirdly, the Antiquities' proclamation of the Imperial regime neglected cultural legacies related to the period since the Era of the Princes (1769–1855). This resulted in the exclusion of cultural values developed from the 18th to the 20th centuries. Moreover, the system of heritage management and its concept were not well developed, which contributed to the gap.

Despite its political disorders due to internal and external factors, the Derg regime attempted to maintain the "meta-narrative" conception to preserve the national identity of the country. However, the process manifested the regime's socialist ideology, which used the narrative to develop it (Marzagora, 2017). The conservation of "progressive" cultural heritage was taken as part of the campaign to create a socialist Ethiopia. Despite its recognition of cultural diversity, the regime's heritagization process, mainly in museum establishments, was criticized for its bias towards ethnic-based ethnographic interests and suspicion of lacking a wider pattern of representations (Aalund, 1985). Before and during the regime, ethnonationalism movements used past narrations and heritage as tools to frame the Imperial regime and the "Great Tradition" as feudalistic and historical causes of inequality. Later on, these movements entered into prolonged military fighting and wiped out the Derg regime in 1991 (Huber, 2021).

11.4 ETHNIC HERITAGIZATION ("MICRO-NARRATIVE") APPROACH OF CULTURAL LEGACIES IN ETHIOPIA: ITS DISCOURSE AND IMPACTS

Ethiopians' memory of the past has remained an object of struggle and contestation in the discourse of defining and redefining the country's past to shape it for the sake of contemporary political interests (Paulos & Getachew, 2005). Cultural heritage is interpreted through an ethnic-based "micro-narrative" perspective (Finneran, 2013). The root of ethnic politics and the misuse of the past can be traced to the colonial era mainly following the Victory of Adwa in 1896. After Adwa, colonial propagators did not welcome Ethiopia's symbolism in the Pan-African movement and its active involvement in the struggle against colonialism (Paulos & Getachew, 2005; Haile, 2017). Different colonial agents such as Newman (1935), Prochazka (1935), and Waugh (1936) attempted to install an ethnic-based categorization among Ethiopians to resume their colonial interest and dismiss the Ethiopianism and Pan-Africanism concepts (Paulos & Getachew, 2005). This scholarship found a place during the "Students Movement" in the 1960s. The students became fascinated by ethnic politics framing the Ethiopian past as a mere construction and worthless. The movement did not merely assume opposition against old regimes. It deeply employed cultural diversity as a grant to promote ethnonationalism, which aggressively used historical narrations as signs of its main agendas: ethnic oppression and marginalization (Messay, 2003; Haile, 2017).

Regardless of its "meta-narrative" orientation, the militaristic nature of the Derg regime intensified ethnic movements and laid the foundation for ethnic-based political fringes (fronts) led by elites claiming Tigrian and Oromo affinities and others who wished to benefit from this political outcome. From the ethnic political perspective, a new conception of Ethiopia different from the previous one has been felt, and both elites have developed their own "national oppression" and "colonial" theses to strengthen their political movements against the "Great Tradition" (Bach, 2014). The liberation fringes gradually strengthened their power and fought the Derg regime for a long period. Finally, the regime was deposed from power and marked the beginning of a new ruling period under the Ethiopian Peoples' Revolutionary Democratic Front (EPRDF) that established an ethnic-based form of government motorized by the Tigray People's Liberation Front (TPLF). Before discussing the discourse and mess of this ethnic-based perspective, it is important to notice its positive contributions. Firstly, the emphasis it pays to intangible heritage enables the government to achieve the inscription of four intangible cultural heritages under UNESCO.[1] Secondly, despite its theoretical policy and misuse, cultural diversity has officially been recognized. Moreover, the system recognizes the active participation of local communities in heritage protection (Huber, 2021) and demonstration of their cultural practices.

However, it has profound negative consequences that are manifested in terms of contestation ideologies, power struggles, degradation of social

[1] https://ich.unesco.org/en/state/ethiopia-ET?info=elements-on-the-lists (Accessed on 19th June 2024).

cohesion or coexistence, ethnic conflict, and negligence of the collective social, cultural, and economic roles of past cultural legacies because its foundation was made predominantly by condemning, disvaluing, vandalizing, abusing, and politically instrumentalizing the past, including the state formation process, regimes, rulers, and cultural legacies (Biniam, 2004; Mackonen, 2008). It heritagized the past as an instrument that fuels ethno-prone conflicts which magnifies its negative consequences (Huber, 2021).

The misuse and misinterpretation of cultural heritage are typically associated with the orchestrated and skewed image of the "rhetoric malware" that disguises the country's past as a mere "prison of nationalities," a recent "invention," and an empire of "Abyssinian colonialists" (Holcomb & Sisai, 1990; Assefa, 2017). In today's political turmoil, competing political factions use heritage and historical narratives as leverage to meet their disparate interests (Huber, 2017). As Finneran (2013) shows, the ethno-political project aims to deconstruct the "meta-narrative" of the country's past that forms the "Great Tradition." It is invested to fracture the common understanding of Ethiopians on their heritage by placing means of cultural exclusions and cultural ignorance. This development involves the devaluation of the country's main built heritages which have national and global outstanding values. Interpretations are merely made from local (ethnic) and political perspectives without considering wider contexts. This approach degrades the sense of collective heritage. Ethno-writers are misinformed about the so-called "meta-narrative" which was developed by considering some cultural heritage values assumed to be representative of the past. The promotion of world heritages such as Aksum, Lalibela, Gondar, and other sites possessing outstanding universal values in the realm of the "Great Tradition" is portrayed negatively, mainly by interpreting such globally represented values only from their local (ethnic) perspective and environment, which is installed to complement ethnic politics (Finneran, 2013).

The scholarship has been propagated in the publications and speeches of ethno-elites, including governmental authorities, political parties, academicians, musicians, activists, and other writers who have commonly agreed with the view that Ethiopia is a colonial output of the Abyssinians. To put this in context, some cases are assessed here. Limbach (2005), in his report concerning the situations during the return of the Aksumite obelisk from Italy to Ethiopia, indicated how cultural heritage was ethnically portrayed creating cultural marginalization. According to his report, the speech made by the then Prime Minister, Meles Zenawi, during the festivity on the return of the obelisk was a kind of disoriented conception of ethno-politicians toward cultural heritage. This speech from a high authority has the potential impact of degrading the Ethiopians' sense of collective heritage. As one of Limbach's interviewees recounted, "Our prime minister tells us that the return of the obelisk is a Tigrian affair, not the business of us southern people." This illustrates how much the speech has influenced people to think of heritage ethnically. This is against the communal nature of heritage as the obelisks have a wider African context and the event during the return had to be taken as a good opportunity to build unity and solidarity among Ethiopians and to foster their economic development (Jara, 2009). In the wider context, that single speech has implications for the devaluation of the long-embedded historical and cultural collective memory and sense of public cultural ownership. This case is an implication of the micro-narrative approach of the government in confining the value of world heritage as mere local property.

Many of the writers invested their effort in discrediting common heroes, exemplary historical events, and national symbols such as the flag that Ethiopians have possessed throughout their history. The devaluation of these aspects is attempted as they have a great connection with the state (re)building process in the 19th century, the destination of ethno-politicians' stories of ethnic oppression and colonial theses (Bach, 2014). Blaming Emperor Menelik II, the architect of Adwa Victory, an icon of the collective historical memory of Ethiopians, is the common project of ethno-writers. Menelik is probably the most accused emperor of modern Ethiopia because of his prominent role in completing the empire formation process and keeping the empire's sovereignty. It is from this point that different ethnic writers employed ethnical cultural redefinition to "liberate" peoples who are retold to be colonized by Menelik.

This scholarship predominantly condemns "Amharization" as a responsible factor for the propagation of Abyssinian "colonialism" and cultural "marginalization" (Holcomb & Sisai, 1990; Asafa, 1993; Abbas, 2014). It culminated in the creation of a number of ethnic political fringes and secessionists, "ethnicstans" or ethnic-based administrations, constitutionalized ethnic politics, and extreme ethnocentrism under the EPRDF (Paulos & Getachew, 2005). The other area subjected to the deconstruction of the Ethiopian past is the country's figurative place in the Pan-Africanism movement. This was initially worked by colonial agents who were unhappy with the global role of Ethiopia. The internal ethnic politics have contributed to weakening the stronghold of Ethiopianism and its role in this global movement against white colonialism, oppression, and racism.

As Blanc & Bridonneau (2016) indicate, the national (and also global) identity of Ethiopian cultural heritage no longer exists because of the extolling of local ethnic identities. The forging of ethnic identities considers the cultures of different people as a closed entity and denies socio-cultural interaction between different peoples and the notion of collective heritage. It merely propagates cultural locality along ethnic lines. It also hampers the promotion of cultural values and their role in nation-building and obstructs the efforts of the country to increase its cultural representation at a global level. Heritage is employed for political purposes to ensure the foundation of ethnic politics. Its economic value is slightly emphasized. This perspective has a problem with the multidimensional values, usage, and administration of cultural heritage (Blanc & Bridonneau, 2016) and this may create a conflict of interest between the government and local communities in administering heritage sites.

The ethnic-based image of the past does not end either in reimagining the country from a "micro-narrative" perspective or rewriting history and reinterpreting cultural values in an ethnic context. It is extended to suit ethnic-based narratives to inflame ethnic-prone conflicts which degrade the collective memory and coexistence of Ethiopians (Assefa, 2017). The ultimate objective of developing ethnic narratives is to see a disintegrated nation, as Asafa (2001) assumes, "Habeshas and Oromos cannot coexist peacefully within a single state system." This signifies the extremity of ethnocentric ideology, which is contrary to the notion of collective heritage. This environment hinders developmental practices (such as tourism) and the "state-building" process and leads to cultural conflicts and ethnic cleansing (Alemseged, 2004; Daniel, 2012).

The ethnic-oriented approach towards the history of Ethiopia has shown change under the regime of Abiy Ahmad since 2018. Initiations were made to promote Ethiopianism and nationally heritagize past legacies (Tsegaye, 2021). Many Ethiopians, both at home and abroad, admired Abiy's attempt to resurrect pan-Ethiopianism (via his philosophy, "Medemer"). However, gradually, it could not escape the influence of the ethno-political mainstream, which demands different political interests, including the dominance of the federal government, maintaining self-administration, or seceding as an alternative. Greedy political interests are continued within the state system, and according to some writers, this creates fear that Ethiopia might face state collapse (Meressa, 2019).

It is illustrative to show here the negative impacts of the ethnic-based discourse. Firstly, it degrades the notion of national collective heritage and hinders further internationalization of the cultural values of the country. Ethiopia is also no longer effective at presenting its heritage in a national context (heritage nationalization). This, in turn, negatively affects the making of heritage at the international level. In comparison to other countries with more than forty recorded world heritage sites, Ethiopia has a small number of registered world heritage sites, despite its rich endowment with past legacies (Huber, 2017). Secondly, the approach negatively affects the role that the country's natural and cultural resources can play in local communities' developmental endeavors. It hinders cultural tourism that can be developed through the proper use of the country's cultural endowments.

Moreover, it misuses cultural diversity and ignites ethnic-prone conflicts that threaten the coexistence of Ethiopians. As different sources show, the conflicts because of the detrimental ethnic politics have resulted in killings, massacres, and displacements of millions of Ethiopians (Abbink, 2006; Lubo, 2012; Bekalu, 2018; Yigzaw & Abitew, 2019). In its worst situation, since November 2020, the country entered into a civil war between the TPLF and the federal government (Tofa, Alagaw & Kinkoh, 2022). This war has destroyed cultural heritage in the Tigray, Amhara, and Afar regions, in addition to its cost of lives. Different governmental and non-governmental reports appeared on the destruction of cultural heritage during the war. The cultural collections of the Dessie Museum (formerly, the palace of Ras Michael of Wollo in the 19th century) in Dessie town were plundered and destroyed by the Tigrayan militants during their occupation of part of the Amhara region.[2] Moreover, reports (such as ICOMOS's statement on the situation in Tigray) were also released on the endangered situation of cultural heritage in Tigray (including the Alnejashi Mosque and the monastery of Debre Damo) due to military attacks and plundering during the conflict.[3] Despite the need for academic investigation to understand the responsible factors and the degree of the impact, these cases show how ethnic politics negatively affects heritagization, heritage management, and cultural tourism in Ethiopia.

11.5 CONCLUSION AND THE WAY FORWARD

This chapter examined the trend of heritagization and interpretation in Ethiopia, with a focus on the ethnic-based perspective of cultural legacies

---

[2] https://www.fanabc.com/english/terrorist-tplf-looted-and-vandalized-dessie-museum (accessed on 19th June 2024).

[3] https: www.icomos.org/images/DOCUMENTS/Secretariat/2021/Statements/ICOMOS_Statement Tigray_EN_final20210304.pdf (accessed on 19th June 2024).

and its numerous negative consequences for Ethiopian unity. However, Ethiopia's past has more prospective opportunities than contradictions to enable its positive role in community development. Four concluding remarks are forwarded to be emphasized by the elites who are dominating the political and socio-economic spaces of the country. Firstly, documenting cultural legacies at national levels has to be given due attention. This will be more important in collective heritabilization. It helps prepare unexplored resources for domestic and international tourism purposes. It also minimizes the spatial coverage disparities of heritagization and the problem of evidence for academic and managerial activities. Currently, as the "Classification of Cultural Heritage into National and Regional Cultural Heritages Proclamation No. 8391/2014" shows, the authority for documentation and management of cultural heritage falls under federal and regional administrative structures. This power-based categorization hinders heritage management and heritagization. It deteriorates the notion of the collective value of heritage at national and global levels. At the regional level, there is a lack of heritage management funds and skilled human resources to properly document and manage the heritage found in the respective regions. Because of such gaps, several significant cultural values are neglected and exposed to destructive anthropogenic and natural factors. A holistic approach to heritage management legislation is needed to enhance heritage protection (Nair, 2016).

Secondly, heritagization has to be made primarily based on a well-defined consensus on the collective values and typologies of cultural resources. All cultural legacies and practices have to be considered part of the common memory of Ethiopians and all humankind. It is a common pride for Ethiopia that the country is "a place of pilgrimage" for several archaeological projects that have reframed the world's perception that once portrayed Africa as a "latecomer" for civilization, as it is found to be a cradle of humankind and culture (Stahl, 2005; Johanson, 2010). Ethiopia possesses varied styles of tangible and intangible cultural values, evidencing cultural development and the continuity of its history. From the records of F. Alvarez in the 16th century to the detailed studies of modern researchers, different kinds of literature testify that Ethiopia is endowed with outstanding past legacies, making it a land of great cultural antiquities and values. Whether it is written or not, any aspect of its history (both glories and agonies) is part of its "Great Tradition" which allows the country to be considered as one of the lands of ancient civilizations (Clapham, 2002). However, this may be challenging due to a lack of sense of belongingness, communal ownership, and freedom to access cultural resources regardless of their location.

The heritagization process needs inclusive consideration of tangible and intangible types of cultures. This helps to avoid the gap created by previous heritagization trends, which favored either tangible or intangible aspects of culture. In the context of Ethiopia, material cultures are either spaces where ritual and secular practices are regularly performed or portable objects through which intangible practices are manifested. It is less likely to see a culture's tangible and intangible characteristics separately because they are found to be intertwined. For instance, the rock-hewn churches of Lalibela embody both the tangible features (architectural, artistic) and the intangible manifestations (spiritual practices and meanings). The nomination of these churches as world heritage was made only by referring to UNESCO's first three criteria, which are associated with outstanding universal values of monuments.[4] Criteria (iv)[5], which is associated with intangible cultural heritage, can be applied to these churches as well as many other cultural heritages throughout the country. The recent inscription of Meskel, Fiche Chamballalla, Geda System, and Timket intangible heritages under the World Heritage List (WHL) deserves appreciation. However, these cultural values are not primarily heritagized as national collective heritage. Moreover, the process of heritagizing additional tangible world heritage is politically sabotaged because of the misperception that associates tangible heritage with the previous "meta-narrative" perspective. This necessitates a rethinking of the previous trend of using an integrative approach to heritaging the two aspects of the cultural past.

Thirdly, revalorizing and reinterpreting the country's cultural past as collective heritage has to be emphasized instead of detrimentally approaching and battling over it. To enable cultural or historical resources to play a positive role, the current ethnic-based approach has to be reoriented into a constructive one emphasizing collectivity. It will probably be challenging to deal with this issue as ethnic politics is continuing in the country. However, as far as the negative impacts are visible, it is advisable to think that collective heritagization and interpretation will have an immense contribution to salvaging Ethiopians' coexistence and supporting the fragile socio-economic development through the employment of cultural resources. More emphasis has to be placed on the non-political values (social, spiritual, economic, aesthetic, technological, architectural, and so on) of heritage. The priority of this valorization fosters the conservation and utilization of heritage for the

[4] See details under https://whc.unesco.org/en/list/18/ (accessed on 19th June 2024).

[5] See its details under https://whc.unesco.org/en/criteria (accessed on 19th June 2024).

sake of community development, as heritage stands for and is owned by the communities as a whole. Interpretation should also be from this perspective.

Finally, the recognition of cultural diversity and converting this blessing into a development resource is an important issue. Ethiopia is a country of different communities with "mosaic" cultural expressions. This diversity is recognized by the federalist government of Ethiopia (MoCT, 2009, 2015). However, the way that this cultural diversity is approached is misguided because of the influence of ethno-politics. The recognition is merely theoretical and politically motivated by the political ideology of the elites. It is well used to install ethnic-based administrative structures based on cultural differences. It is also well used as a springboard to address the interests of different individuals or groups. Therefore, it is likely to be said that cultural diversity is primarily recognized for the political purposes of the state. As a result, the country's diverse form of culture is not well used to enable developmental practices such as tourism. Appreciating and recognizing living cultural practices is the foundation for recognizing different cultural practices and fostering learning, cultural tolerance, and harmonious relationships among various communities. Apart from recognition, cultural diversity has to be converted into a valuable resource for development.

KEYWORDS

- **African Union**
- **cultural legacies**
- **Ethiopia**
- **ethno-politics**
- **heritagization**
- **interpretation**
- **Organization of African Unity**
- **World Heritage List**

REFERENCES

Aalund, F. (1985). *Master Plan for the Preservation and Presentation of Cultural Heritage: Ethiopia*. Paris: UNESCO.

Abbay, A. (2004). Diversity and state-building in Ethiopia. *African Affairs, 103*(413), 593–614.

Abbink, J. (2006). Ethnicity and conflict generation in Ethiopia: Some problems and prospects of Ethno-regional federalism. *Journal of Contemporary African Studies, 24*(3), 389–413.

Aikawa-Faure, N. (2009). From the proclamation of masterpieces to the convention for the safeguarding of intangible cultural heritage. In L. Smith & N. Akagawa (Eds.), *Intangible Heritage* (pp. 13–44). London: Routledge.

Albert, M., Bernecker, R., & Rudolff, B. (2013). *Heritage Studies: Perspectives in Heritage Studies*. Berlin: Walter de Gruyter GmbH.

Alvarez, F. (1881). *Narrative of the Portuguese Embassy to Abyssinia: During the Years 1520–1527* (trans. Lord Stanley of Alderley). London: The Hakluyt Society.

Ashworth, G. J., Graham, B., & Tunbridge, J. E. (2007). *Pluralizing Pasts: Heritage, Identity and Place in Multicultural Societies*. London: Pluto Press.

Atnafu, B. (2018). Ethnic cleansing in Ethiopia. *The Canadian Journal of Peace and Conflict Studies, 50*(1), 77–104.

Avrami, E., Mason, R., & de la Torre, M. (2000). *Values and Heritage Conservation: Research Report*. Los Angeles: The Getty Conservation Institute.

Bach, J. (2014). EPRDF's nation-building: Tinkering with convictions and pragmatism. *Cadernos de Estudos Africanos, 27*, 103–126.

Blanc, G., & Bridonneau, M. (2016). Making heritage in Ethiopia/Faire le patrimoine en Éthiopie. *Annales d'Ethiopie, 31*, 13–23.

Buxton, D. (1947). The Christian antiquities of northern Ethiopia. *Archaeologia, 92*, 1–42.

Chhabra, D. (2010). *Sustainable Marketing of Cultural and Heritage Tourism*. London: Routledge.

Clapham, C. (2002). Rewriting Ethiopian history. *Annales d'Ethiopie, 18*, 37–54.

Cohen, D. (2001). Cultural variation: Considerations and implications. *Psychological Bulletin, 127*(4), 451–471.

Deacon, J. (2014). An overview of the history of the nomination of cultural World Heritage Sites in Africa. In S. Makuvaza (Ed.), *The Management of Cultural World Heritage Sites and Development in Africa: History, Nomination Processes and Representation on the World Heritage List* (pp. 1–16). New York: Springer.

During, R. (2011). European heritage discourses, a matter of identity construction? In R. During (Ed.), *Cultural Heritage and Identity Politics* (pp. 17–30). Silk Road Research Foundation.

Finneran, N. (2007). *The Archaeology of Ethiopia*. London: Routledge.

Finneran, N. (2013). Lucy to Lalibela: Heritage and identity in Ethiopia in the twenty-first century. *International Journal of Heritage Studies, 19*(1), 41–61.

Galaty, M. L., & Watkinson, C. (2004). The practice of archaeology under dictatorship. In M. L. Galaty & C. Watkinson (Eds.), *Archaeology under Dictatorship* (pp. 1–18). USA: Springer.

Gebrewahd, M. T. (2019). Nation-building predicament, transition fatigue, and fear of state collapse: An emerging phenomenon in post-2015 Ethiopia. *Hungarian Journal of African Studies, 13*(5), 32–50.

Girma, A. (2018). Archaeology, politics and nationalism in nineteenth- and early twentieth-century Ethiopia: The use of archaeology to consolidate monarchical power. *Azania: Archaeological Research in Africa, 53*(3), 398–416.

Gnamo, A. H. (2014). *Conquest and Resistance in the Ethiopian Empire, 1880–1974: The Case of Arsi Oromo*. Leiden: Brill.

Gosden, C., Cunliffe, B., & Joyce, R. A. (2009). *The Oxford Handbook of Archaeology*. Oxford: Oxford University Press.

Haile-Mariam, J. (2009). The cultural benefits of the return of the Axum obelisk. *Museum International, 61*(1–2), 48–51.

Harrison, R., Bartolini, N., Desilvey, C., Holtorf, C., Lyons, A., Macdonald, S., May, S., Morgan, J., & Penrose, S. (2020). *Heritage Futures: Comparative Approaches to Natural and Cultural Heritage Practices*. London: UCL Press.

Harvey, D. C. (2008). The history of heritage. In B. Graham & P. Howard (Eds.), *The Ashgate Research Companion to Heritage and Identity* (pp. 19–36). England: Ashgate Publishing.

Hassner, P. (2009). Beyond politics and history: The need for conceptual and ethical dialogue. In H. Swoboda & J. M. Wiersma (Eds.), *Politics of the Past: The Use and Abuse of the History* (pp. 15–28). The Socialist Group in European Parliament.

Henze, P. B. (2000). *Layers of Time: A History of Ethiopia*. New York: Palgrave.

Holcomb, B. K., & Ibbsa, S. (1990). *The Invention of Ethiopia*. Trenton: The Red Sea Press.

Howard, P. (2003). *Heritage Management, Interpretation, Identity*. London & New York: Continuum.

Howard, S. (2021). *Ethiopia: The Essential Guide to Customs and Culture*. Britain: Kuperard.

Huber, M. (2017). *Ethiopia and the Beginnings of the UNESCO World Heritage Programme 1960–1980* (PhD dissertation). Humboldt University.

Huber, M. (2021). *Developing Heritage-Developing Countries: Ethiopian Nation-Building and the Origins of UNESCO World Heritage, 1960–1980*. Berlin: Walter de Gruyter GmbH.

ICOMOS (1999). *Cultural Tourism Charter*. Mexico: ICOMOS.

ICOMOS. (2008). *The ICOMOS Charter for the Interpretation and Presentation of Cultural Heritage Sites*. Quebec: ICOMOS.

Jalata, A. (1993). Sociocultural origins of the Oromo national movement in Ethiopia. *Journal of Political and Military Sociology, 21*, 267–286.

Jalata, A. (2001). *Fighting Against the Injustice of the State and Globalization: Comparing the African American and Oromo Movements*. New York: Palgrave.

Jamieson, W. (1998). Cultural heritage tourism planning and development: Defining the field and its challenges. *APT Bulletin: The Journal of Preservation Technology, 29*(3–4), 65–67.

Johanson, D. C. (2010). From Lucy to language: Our place in nature. *Journal of the Siena Academy of Sciences, 2*, 8–9.

Joussaume, R. (2010). Təya. In S. Uhlig (Ed.), *Encyclopedia Aethiopica* (pp. 939–940). Wiesbaden: Harrassowitz Verlag.

Kasomo, D. (2012). An assessment of ethnic conflict and its challenges today. *African Journal of Political Science and International Relations, 6*(1), 1–7.

Kebede, M. (2003). Eurocentrism and Ethiopian historiography: Deconstructing semitization. *International Journal of Ethiopian Studies, 1*(1), 1–19.

Kottak, C. P. (2015). *Cultural Anthropology: Appreciating Cultural Diversity* (16th ed.). USA: McGraw-Hill.

Larebo, H. (2017). Lemehonu yebiher bihereseb chikona beEthiopia neber woy? *International Journal of Ethiopian Studies, 11*(1–2), 1–44.

Limbach, I. (2005). The Axum obelisk returns, but some still grumble. *Archaeology, 58*(4), 10–11.

Marcus, H. G. (1994). *A History of Ethiopia*. Berkeley: University of California Press.

Markakis, J. (1989). Nationalities and the state in Ethiopia. *Third World Quarterly, 11*(4), 118–130.

Marzagora, S. (2017). History in twentieth-century Ethiopia: The 'Great Tradition' and the counter-histories of national failure. *Journal of African History, 58*(3), 425–444.

Mason, R. (2002). Assessing values in conservation planning: Methodological issues and choices. In M. de la Torre (Ed.), *Assessing the Values of Cultural Heritage* (pp. 5–30). The J. Paul Getty Trust.

McPherron, S. P., Alemseged, Z., Curtis, W. M., Wynn, J. G., Read, D., Geraads, D., Bobe, R., & Béarat, H. A. (2010). Evidence for Stone-tool-assisted consumption of animal tissues before 3.39 million years ago at Dikika, Ethiopia. *Nature, 466*(12), 857–860.

Mehretu, A. (2017). Delegitimization of the collective identity of Ethiopianism. *International Journal of Ethiopian Studies, 11*(1–2), 45–70.

Merryman, J. H. (1986). Two ways of thinking about cultural property. *The American Journal of International Law, 80*(4), 831–853.

Milkias, P., & Metaferia, G. (Eds.). (2005). *The Battle of Adwa: Reflections on Ethiopia's Historic Victory against European Colonialism*. New York: Algora Publishing.

Ministry of Culture and Tourism (MoCT). (2009). *Tourism Development Policy*. Addis Ababa: Ministry of Culture and Tourism.

Ministry of Culture and Tourism. (2015). *Sustainable Tourism Master Plan, 2015–2025*. Addis Ababa: United Nations Economic Commission for Africa.

Munro-Hay, S. (1991). *Aksum: An African Civilization of Late Antiquity*. Edinburgh: Edinburgh University Press.

Nair, V. S. (2016). Perceptions, legislation, and management of cultural heritage in Ethiopia. *International Journal of Cultural Property, 23*, 99–114.

Newman, E. W. P. (1935). Abyssinia. *RUSI Journal, 80*(520), 842–849.

Nilsson, P. Å. (2018). Impact of cultural heritage on tourists: The heritagization process. *Athens Journal of Tourism, 5*(1), 35–54.

Nowacki, M. (2012). *Heritage Interpretation*. Poznań: Akademia Wychowania Fizycznego.

Nowacki, M. (2021). Heritage interpretation and sustainable development: A systematic literature review. *Sustainability, 13*(4383), 1–16.

Pankhurst, R. (1973). The library of emperor Tewodros II at Mäqdäla (Magdala). *Bulletin of the School of Oriental and African Studies, 36*(1), 15–42.

Pankhurst, R. K. (1982). The history of the Kwer'ata Re'esu: An Ethiopian icon. *African Affairs, 81*(322), 117–125.

Phillipson, D. (2009). *Ancient Churches of Ethiopia: Fourth-fourteenth Centuries*. New Haven and London: Yale University Press.

Prochazka, B. R. (1935). *Abyssinia: The Powder Barrel*. London: British International News Agency.

Silberman, N. A. (2013). Heritage interpretation as public discourse: Towards a new paradigm. In M. Albert, R. Bernecker, & B. Rudolff (Eds.), *Understanding Heritage: Perspectives in Heritage Studies* (pp. 21–34). Berlin: Walter de Gruyter GmbH.

Sjöholm, J. (2016). Heritagisation, re-heritagisation and de-heritagisation of built environments: The urban transformation of Kiruna, Sweden. (Doctoral dissertation, Luleå University of Technology).

Smith, L. (2004). *Archaeological Theory and the Politics of Cultural Heritage*. London: Routledge.

Srivastava, S. (2019). *Conservation and Promotion of Heritage Tourism*. USA: IGI Global.

Stahl, A. B. (Ed.). (2005). *African Archaeology: A Critical Introduction*. USA: Blackwell Publishing.

Teferi, L. (2012). The post-1991 'inter-ethnic' conflicts in Ethiopia: An investigation. *Journal of Law and Conflict Resolution, 4*(4), 62–69.

Timothy, D. J., & Nyaupane, G. P. (Eds.). (2009). *Cultural Heritage and Tourism in the Developing World: A Regional Perspective*. London: Routledge.

Tofa, M., Alagaw, A., & Kinkoh, H. (2022). *Political and Media Analysis on the Tigray Conflict in Ethiopia*. Brussels: European Institute of Peace.

Trigger, B. G. (1996). *A History of Archaeological Thought* (2nd ed.). Cambridge: Cambridge University Press.

Tsegaye Ebabey. (2021). Ethno-nationalists' anachronistic tendencies in deconstructing the Ethiopian past and the role of the past in unifying the people. *ERJSSH, 8*(2), 105–128.

Tunbridge, J. E., & Ashworth, G. J. (1996). *Dissonant Heritage: The Management of the Past as a Resource in Conflict*. England: John Wiley and Sons.

UNESCO. (2003). *Convention for the Safeguarding of the Intangible Cultural Heritage* (32nd Session of the General Conference). Paris: UNESCO.

UNESCO. (2015). *Basic Texts of the 2005 Convention on the Protection and Promotion of the Diversity of Cultural Expressions*. Paris: UNESCO.

Wagner, E. (2005). Harär city structure and main buildings. In S. Uhlig (Ed.), *Encyclopedia Aethiopica* (pp. 1013–1015). Wiesbaden: Harrassowitz Verlag.

Waugh, E. (1936). *Waugh in Abyssinia*. London: Longmans.

Weldegebreil, B. (2004). Memories of the victory of Adwa: A focus on its commemoration, 1941–1999 (Master's thesis, Addis Ababa University).

Worthing, W. D., & Bond, S. (2008). *Managing Built Heritage: The Role of Cultural Significance*. Oxford: Blackwell Publishing.

Yigzaw Gedifew & Abitew Endalsasa (2019). Causes and impacts of internal displacement in Ethiopia. *African Journal of Social Work, 9*(2), 32–41.

CHAPTER 12

STAKEHOLDER COLLABORATION FOR SUSTAINABLE TOURISM DEVELOPMENT AT CHAM COMMUNITY'S LIVING-SACRED SITES IN VIETNAM: A LIVING HERITAGE APPROACH

TUYEN DAI QUANG

Faculty of Tourism and Vietnamese Studies, Nguyen Tat Thanh University, Vietnam

ABSTRACT

Vietnam has been promoting tourism development, especially in ethnic minority areas like the Cham community, which possess significant cultural tourism resources. However, tourism development based on Cham cultural heritage resources faces challenges, and this study explores collaboration between the Cham community and the local government to propose a sustainable management framework for living-sacred sites. The study, based on qualitative research, found that despite the Vietnamese State's efforts to preserve and promote Cham cultural heritage, collaborations between the Cham community and the local government remain weak due to miscommunication and a lack of trust. The proposed management framework requires a collaborative approach involving both the government and the community in decision-making processes to ensure culturally sensitive and sustainable tourism development. This study highlights the importance of establishing

Dynamics of Heritage Management and Tourism Development: Trends, Practices, and Approaches.
Surabhi Srivastava & Ruchika Kulshrestha (Eds.)

effective collaborations between the Cham community and the local government for the sustainable development of tourism in the Cham community's living-sacred sites. To promote culturally sensitive and sustainable tourism development, the suggested management framework underscores the need for including local participation in decision-making processes. The study's findings contribute to the ongoing dialogue on heritage conservation and tourism development, particularly in ethnic minority areas such as the Cham community.

12.1 INTRODUCTION

There has been considerable academic and applied focus on the dynamics of heritage management and tourism development in recent periods of time (Mason, 2015; Silva & Chapagain, 2013; Timothy & Nyaupane, 2009). This growing interest is due to the increasing recognition of the potential of cultural heritage in promoting sustainable tourism development (STD), especially in ethnic minority areas (Tilley, 1997; Yang & Wall, 2009). As a result, a growing body of literature has emerged to explore the trends, practices, and approaches in heritage management and tourism development (Bradford & Lee, 2004; Garrod & Fyall, 2000; Silva & Chapagain, 2013). The notion of STD, which underlines the need to harmonize socioeconomic, social, and environmental components of tourism, has become an indispensable foundation for comprehending the intricate relationship between heritage governance and tourism progress (Lu & Nepal, 2009; UNWTO, 2015). Over the past few decades, STD has gained increasing attention as a means of promoting economic growth, particularly in ethnic communities (Bruner, 2005; Buckley, 2012; Li & Hinch, 1997). Heritage conservation and tourism development have been identified as potential vehicles for such growth. In Vietnam, like many other countries, various tourism policies have been implemented to promote tourism development, including in ethnic minority areas. However, tourism development in such areas presents challenges, particularly in terms of cultural heritage conservation. Stakeholder collaboration has been identified as an important aspect of the development of sustainable tourism that ensures heritage conservation and economic growth (Byrd, 2007; Dabphet, 2016).

The Cham community in Vietnam is a case in point. The community has rich cultural tourism resources that have made it an emerging tourism destination in the South-Central region of Vietnam. However, the exploitation of Cham cultural heritage resources for tourism activities has presented challenges. The Cham people's living-sacred sites are significant religious and cultural symbols that require a living heritage perspective to ensure their sustainability for future generations (Tuyen D. Quang, Noseworthy, & Paulson, 2022).

Collaboration between stakeholders is essential to address the challenges associated with tourism development in ethnic minority areas. These stakeholders include the local government, tourism developers, heritage experts, and the local community. Collaborative efforts can help to develop a framework for sustainable tourism development and cultural heritage conservation (Byrd, 2007; Dabphet, 2016).

In the context of cultural tourism development, the living heritage strategy is vital, especially in ethnic minority communities such as the Cham people in Vietnam. This approach emphasizes the importance of involving the local community in the management and conservation of cultural heritage, recognizing their roles as creators and transmitters of cultural values and practices (Lenzerini, 2011; Wijesuriya, 2015). By taking a living heritage approach, tourism development can be managed in a way that respects cultural diversity, supports local livelihoods, and preserves the cultural heritage for future generations (Lenzerini, 2011; Wijesuriya, 2015). Therefore, this study's proposed management framework takes a living heritage approach to ensure STD and cultural heritage conservation in the living-sacred sites of the Cham people (Gürçayir Teke, 2018; Tuyen D. Quang et al., 2022).

This study focuses on the Cham ethnic group in Vietnam and explores the perceptions related to collaboration activities between the Cham community and the local government. The study aims to propose an appropriate management framework to develop tourism in the living-sacred sites of the Cham community in a sustainable way. The study utilizes a living heritage approach and a qualitative research method to explore insightful perspectives from each stakeholder and identify issues and challenges in current collaborations. The findings highlight the significance of stakeholder collaboration for sustainable heritage management and provide a management framework for managing living sacred sites in ethnic minority areas in Vietnam.

12.2 LITERATURE REVIEW

12.2.1 LIVING HERITAGE APPROACH

Living heritage, which is rooted in the past, is continuously adapting to contemporary conditions by embracing change (Bui, Jones, Weaver, & Le,

2020; Poulios, 2014; Walter, 2021). Central to numerous cultural traditions is the continuity between the historical and contemporary eras (Poulios, 2014; Wijesuriya, 2007). It is essential to integrate local people in the conservation and administration of ancient sites. This is underscored by the living heritage concept, which encourages a people-centered strategy for preservation that benefits both ancient sites and society (Poulios, 2014; Sarah & Wijesuriya, 2014). The involvement of the community is crucial in managing and determining the future of living heritage (Renault & Collange, 2008). According to Weise (2013), living heritage represents "a community's soul and its unique cultural identity." Thus, any modification to the governance of cultural assets must respect the local culture as defined by locals (Poulios, 2014; Weise, 2013).

Effective conservation practices must recognize the significance of local people's input in managing living heritage. Poulios (2014) suggests a variety of standards for identifying and managing living heritage, including the continuity of the heritage site's original function, the continuity of the community's contact with the heritage site, the continuity of engaging in tangible and intangible expressions to accommodate shifting conditions, and the continuity of the community's care for the heritage site via monitoring systems, traditional knowledge, and preservation practices. This approach considers heritage as an intrinsic component of contemporary communal life (Bui et al., 2020; Poulios, 2014; Walter, 2021).

Disregarding local engagement and continuity might have unintended detrimental effects on the value of heritage to local people (Poulios, 2014; Weise, 2013; Wijesuriya, 2014). When properly administered, a community's heritage assets may endure to provide spiritual, sociocultural, and financial consequences (Poulios, 2014; Weise, 2013; Wijesuriya, 2014). Therefore, the dynamic and evolving nature of living heritage reflects the changing needs and practices of the communities that practice it, and often, communities perform their duties under the supervision of heritage bodies that govern the sites (Poulios, 2014). Heritage belongs to the community members whose traditions are represented in the notion of heritage, and in the recent two decades, the preservation progression has established programs and activity centers to convey this (Miura, 2005; Poulios, 2014; Walter, 2021).

12.2.2 STAKEHOLDER COLLABORATIONS

Stakeholder collaboration is recognized as a crucial element in heritage conservation and sustainable tourism development, particularly in ethnic minority areas. It has been emphasized by scholars that stakeholder collaboration is crucial in ensuring the effective management of heritage sites and preserving cultural values (Aas, Ladkin, & Fletcher, 2005; Dabphet, 2016). Local communities are the main custodians of cultural heritage, thus including them as crucial partners in the process of heritage conservation and tourism development is very important (Aas et al., 2005; Byrd, 2007; Dabphet, 2016).

Various forms of collaboration have been employed, such as co-management, participatory planning, and community-based tourism (Balen & Vandesande, 2015; Mak, Cheung, & Hui, 2017). Co-management of cultural sites includes numerous actors, including community members, authorities, and tour companies (Borrini-Feyerabend, et al., 2007). Participatory planning engages local groups actively in the planning and decision-making process for tourism development and heritage conservation, while community-based tourism aims to promote sustainable tourism by involving local people in tourism development and management (Balen & Vandesande, 2015; Mak et al., 2017; Yanes, Zielinski, Diaz Cano, & Kim, 2019). However, challenges such as conflicting interests, power dynamics, and limited resources can affect the effectiveness of collaboration among stakeholders (Buckley, 2012; Dong & Nguyen, 2022). Moreover, a lack of communication and trust among stakeholders might impede the successful implementation of collaborative initiatives (Balen & Vandesande, 2015; Mak et al., 2017). Therefore, effective stakeholder collaboration requires careful consideration of various factors and the use of appropriate collaboration strategies.

To guarantee the sustainable use and preservation of cultural heritage, ethnic and indigenous groups have an essential position in tourism development (Aas et al., 2005; Byrd, 2007; Dabphet, 2016; Timothy & Nyaupane, 2009). The participation of these communities may foster a sense of belonging and glory, leading to the sustainable consolidation of cultural preservation and tourism progression (Mason, 2015; Timothy & Nyaupane, 2009). To achieve this, collaboration among various stakeholders is crucial, including local government, tourism developers, heritage experts, and the local community (Dabphet, 2016; Heslinga, Groote, & Vanclay, 2019).

In ethnic or indigenous heritage contexts, power imbalances among stakeholders can lead to the marginalization of the host community and a lack of participation in decision-making (Carr, Ruhanen, & Whitford, 2016a; Sharpley, 2000). Therefore, it is essential to recognize and respect the cultural values and beliefs of the locals to promote effective collaborations (Aas et al., 2005; Tuyen D. Quang et al., 2022).

Effective collaboration requires the roles of different stakeholders to be clearly defined, including the host community, government, tourism businesses, non-governmental organizations, and tourists (Aas et al., 2005; Dabphet, 2016). Collaboration provides opportunities for addressing power imbalances and ensuring the inclusion of the host community in decision-making processes (Balen & Vandesande, 2015; Mak et al., 2017; Yanes et al., 2019). However, challenges such as cultural differences, conflicting interests, and the investment required in building relationships and trust between stakeholders must be overcome (Bramwell, 2010; Langfield, Logan, & Nic Craith, 2010).

12.2.3 EFFECTS OF POWERS IN COLLABORATIONS

The connection between power and STD is a topic of interest in tourism research (Dolezal & Novelli, 2022; Dong & Nguyen, 2022; Nunkoo, 2017). Power dynamics in collaborations between ethnic/indigenous communities and tourism stakeholders have been investigated in several studies. For instance, the study by Alexis Bunten (2011) examined the power dynamics in the collaboration between four indigenous communities and tourism businesses in Alaska, while the study by Heldt & Miranda (2021) investigated the power dynamics in the collaboration between Indigenous communities and tourism stakeholders in Québec, Canada. These studies found that there were power imbalances, with the indigenous communities having limited power in decision-making processes, despite being an integral part of the tourism product.

Numerous stakeholders, including authorities, non-governmental organizations, tourism enterprises, local people, visitors, researchers, and the media, may collaborate and make a significant contribution to the growth of destinations (Aas et al., 2005). The origins and forms of stakeholder influence are determined by the resources they possess and their network relationships. In terms of resources and political positions, governments and large corporations are usually seen as more influential stakeholders than community members. Therefore, it is essential to realize how less-powerful actors wield power over influential actors and compel them to include the interests of less-powerful stakeholders in tourism development (Dolezal & Novelli, 2022; Saito & Ruhanen, 2017).

Community involvement is crucial to achieving sustainable development of tourist destinations. Community involvement can enhance social capital, improve destination competitiveness, and contribute to the long-term sustainability of tourism (Dolezal & Novelli, 2022; Jamal & Stronza, 2009). However, power imbalances between stakeholders can limit community involvement, leading to marginalization and exclusion of local communities from decision-making processes (Dolezal & Novelli, 2022; Saito & Ruhanen, 2017). According to Bornhorst et al. (2010), large companies and governments are often seen as more powerful stakeholders, while the local community is considered less powerful and excluded from decision-making. Thus, power imbalances can cause tension and undermine the STD (Dolezal & Novelli, 2022; Dong & Nguyen, 2022; Saito & Ruhanen, 2017).

In sum, power dynamics and local participation are crucial for achieving sustainable tourism development in ethnic/indigenous communities. By addressing power imbalances and involving local communities in decision-making processes, tourism may promote economic growth, social capital, and cultural preservation while limiting negative effects.

12.3 RESEARCH METHODS

12.3.1 METHODS AND DATA COLLECTION

This study used a qualitative case study methodology to investigate the power dynamics and stakeholder collaborations in tourism development at the Cham heritage sites. Case studies are commonly used in power-based tourism studies to provide a comprehensive and in-depth understanding of stakeholder relationships (Cornelisse, 2020). Data and information were collected over a 12-month period from 2016 to 2017, exploring the views, experiences, and opinions of people on particular topics utilizing a multidimensional method (Bernard, 2011; Creswell, 2014).

I did ethnographic fieldwork and on-site observations to obtain insight into cultural activities, religious views, and behaviors, and to comprehend contacts between various members and local opinions of minority groups who may speak distinct tongues (Dwyer, Gill, & Seetaram, 2012). The on-site observations helped the author to understand the study site, capture an overview of current tourism, and select potential participants. In addition, semi-structured interviews were used to explicate individuals' sentiments and viewpoints about their daily experiences (Patton, 2002). The interview's objective was to investigate the collaborations between the Cham community and local government in tourism development at the Cham heritage sites. This enabled me to obtain a better comprehension of social phenomena than using quantitative techniques (Creswell, 2014). The authors employed a purposive and convenient sampling approach to select 51 participants who were living in or were familiar with tourism development at the study site. I was using a convenient and purposeful sample technique to identify 51 individuals who resided in or were associated with tourism growth at the research location. The technique of purposive sampling allowed for the selection of respondents with relevant expertise who might offer essential data about Cham culture and involvement at the research location (Creswell, 2014; Patton, 2002).

The author conducted formal interviews with six government officials. The perspectives of Vietnamese officials were similar, and their views on Cham history and culture were relatively perfunctory, so the author did not pursue a deeper line of questioning with them. The structured portion of the interviews focused on basic information before delving into more general questions about Cham community participation in heritage tourism development and collaboration with local authorities. Cham people had diverse perspectives on the nuances of Cham history and culture.

The interviews comprised three parts: (1) demographic information; (2) challenges in stakeholder collaborations (Buckley, 2012; Timothy & Ron, 2013); and (3) their expectations in collaboration. In each section, we also asked for recommendations to improve stakeholder collaboration in the region.

The research categorized the individuals interviewed and assigned their corresponding codes. The categories included Cham scholars, Cham priests, Cham older intellectuals, Cham youth, heritage authorities, and Cham women. The research team conducted both formal interviews and informal conversations with individuals in each category. In total, there were 71 interviews, with 31 being formal interviews and 40 being informal conversations.

Interviews in the study lasted 45–90 minutes and were conducted in Cham or Vietnamese. The author's insider-outsider status provided observations on Cham community members' views on conservation at sacred sites. Qualitative content analysis and coding were used to analyze the transcribed data. Quotes were extracted using respondent codes to ensure anonymity.

12.3.2 RESEARCH CONTEXT

The Cham people have a long and rich cultural history originating from Vietnam's central region. They are the earliest possessors of a written language in Southeast Asia and their cultural practices and beliefs have significantly enriched the Vietnamese national identity. The Cham communities in Ninh Thuận province represent the oldest Indigenous Cham settlements in Southeast Asia, with their cultural heritage still intact. In 2019, there were 178,948 Cham people in Vietnam, and 12.39% of Ninh Thuận province's population are Cham (General Statistics Office, 2019). The Cham people have a distinct cultural richness that is evident in Ninh Thuận through their traditional festivals, customs, and beliefs, which are expressed in their writing, costumes, architecture, sculpture, and traditional crafts (Bien, An, & Dop, 1991; Tuyen, 2019).

Ninh Thuan Province, situated on Vietnam's South-Central coast, was historically a part of the Panduranga subregion of the ancient Champa kingdom, the last Cham territory to be seized by Vietnamese authorities in 1832 (Dharma, 2012). As a result, the Cham community in Ninh Thuan has been able to preserve its cultural heritage, including Cham manuscripts, traditional arts, rituals, festivals, ceremonies, language, and literature, making it a prime location for the study of Cham cultural heritage and history (Abdul, 2013).

This study focuses on the Cham people of Ninh Thuận, Vietnam, a long-standing ethnic group with a rich history and cultural tradition originating from the central region of the country. They are the oldest Indigenous Cham settlements in Southeast Asia and still preserve their cultural heritage through distinct customs, traditional festivals, and ancient beliefs, which are expressed in their writing, costumes, architecture, sculpture, and traditional crafts. The Cham people are actively involved in over 70 different festivals and religious practices throughout the year, with religion playing a central role in their socio-economic and cultural activities. These festivals and ceremonies are led by religious dignitaries, and the Cham community actively participates in all of them (Sakaya, 2003).

The government of Vietnam has recognized and promoted the Cham people's traditional approaches to the custodianship of sacred sites. In 2012, representatives of the Ahier Cham community were appointed to consult with government authorities on cultural and religious issues. This is a significant step in preserving the tangible cultural heritage of the Cham people. In particular, the temple-tower architecture system scattered throughout the central region and in Ninh Thuận province reflects the history, religion, architecture, and sculpture of the Cham people (Figure 12.1).

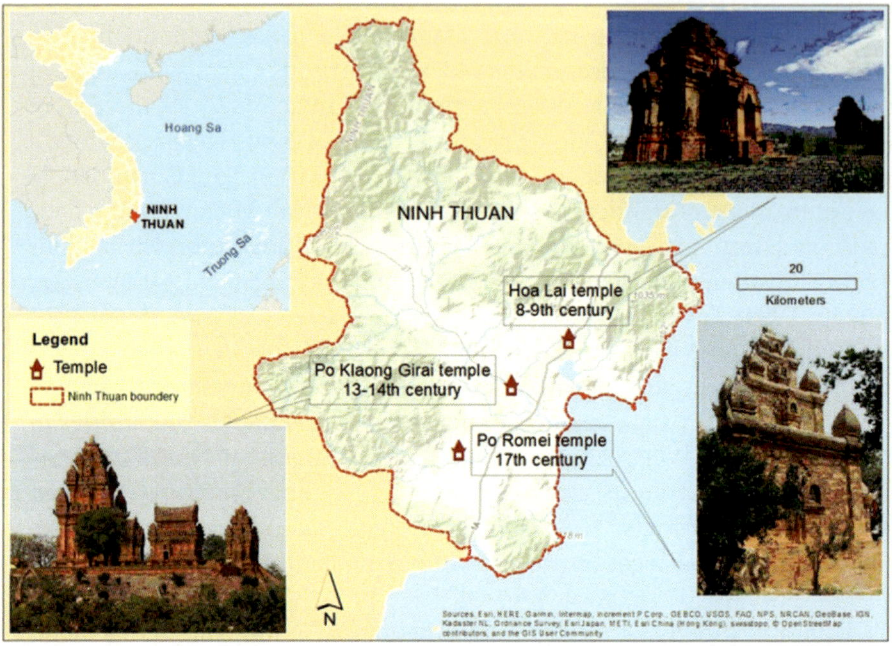

FIGURE 12.1 Map of the Cham temples in Ninh Thuan province (Quang Tuyen Dai, 2022).

The Ninh Thuận Cham society is divided into two main classes: the Halau Janâng, which consists of the priestly system, and the Gihéh, which is made up of the laity. The Cham custodians have the responsibility of overseeing and safeguarding the natural resources and sacred sites, including temples, shrines, and Kut graves. Additionally, some individuals provide assistance to the custodians in the management of these sites. The Cham people's matriarchal customs are still preserved, and many forms of Cham cultural heritage are recognized by the state. Ninh Thuận province's cultural richness attracts tourists and has contributed positively to the region's socioeconomic development, but negative issues such as conflicts have also emerged (Tuyen D. Quang et al., 2022).

In 2019, Ninh Thuận received a total of 2.35 million visitors, a 7.3% increase from the previous year. Of these visitors, 100,000 were international tourists, representing a 25% increase from the previous year, while 2.25 million were domestic tourists, with an annual increase of 6.6% (Sở VH-TT-DL, 2020). Most of the international tourists visiting Ninh Thuận are day-trippers, mostly connecting to the tourism route from Nha Trang (Khánh Hòa). Domestic tourists are the main source of tourism for the Ninh Thuận market, typically accounting for 92–96% of the total number of visitors. From 2011 to 2019, domestic tourist numbers increased at a compound annual growth rate (CAGR) of 14.6% per year. Between 2015 and 2019, the number of both international and domestic tourists had CAGRs of 25.7% and 11.4% per year, respectively. However, in 2015, the number of international tourists decreased by approximately 57.9% compared to the same period. This was a time when the number of Russian visitors to Vietnam significantly dropped due to the depreciation of the Russian ruble, causing ripple effects for beach tourism markets like Ninh Thuận, Khánh Hòa, & Bình Thuận (Figure 12.2).

FIGURE 12.2 Number of Visitors to Ninh Thuan Province from 2019–2023.

Notes: Data compiled by author from Ninh Thuận Department of Culture, Sports, and Tourism reports.

The Cham culture plays an essential role in attracting tourists to Ninh Thuận and has become a tourist product in Vietnam. Although the rise in tourism has positively contributed to the region's socioeconomic development, negative issues have also emerged (Quang et al., 2022). Despite changes throughout their history, preserving this heritage is crucial to maintaining their distinct cultural identity. To ensure the preservation and sustainability of their heritage, it is imperative to manage and protect the natural resources, religion, and culture effectively. Given the importance of tourism to the region, collaboration between stakeholders, especially those related to ethnic or indigenous heritage and tourism development, is necessary. Such collaboration can provide valuable insights into opportunities and challenges that can help promote sustainable tourism and preserve Cham's cultural heritage in Ninh Thuan.

12.3.3 THE IDENTIFICATION OF STAKEHOLDERS

Identifying stakeholders and involving various parties with a shared interest in sustainability are crucial steps towards achieving mutual and beneficial development goals, not only in Vietnam but also globally. In this study, stakeholders refer to those directly impacted by conservation and tourism developments at Cham temples. Multiple groups from various sectors are involved in utilizing cultural heritage resources for STD, making it necessary to clearly define the roles each stakeholder plays, despite the inherent difficulties. In the case of Cham temples, there are various stakeholders, including key stakeholders and secondary stakeholders.

In this study, stakeholders are categorized into two groups: primary and indirect. The primary stakeholders include members of the Cham community who are directly involved in spiritual activities, as well as state and local heritage authorities responsible for the conservation of heritage and the development of tourism at the site. The indirect stakeholders, on the other hand, are those who are not directly involved in spiritual or economic activities but can still be affected by or have an influence on them. This includes the local Kinh community living in the areas surrounding the temple, local businesses selling tourism products related to the temple, and tourist enterprises and companies that bring tourists to the temple.

To evaluate the effectiveness of existing management strategies with respect to the needs of the Cham community, I conducted research to identify the stakeholders and examine their roles in heritage management. I gave priority to the voices of primary stakeholders, particularly those of the Cham community, over exogenous actors, to contrast their views with the existing strategies of those with the most power in preserving Cham cultural heritage sites. This approach allowed me to assess the effectiveness of the current management strategies and their alignment with the needs of the Cham community.

12.4 KEY FINDINGS FROM FIELD RESEARCH ON TOURISM DEVELOPMENT AND STAKEHOLDER COLLABORATION AT CHAM SACRED SITES

12.4.1 EMERGING CHALLENGES IN TOURISM DEVELOPMENT AT THE CHAM SACRED SITES

Based on our survey results, it is evident that there are significant negative issues arising from the exploitation of the sacred temple heritage of the Cham people for tourism development. The public and social media often have intense discussions surrounding these issues. One of the most significant problems cited is the lack of engagement with the Cham population, which has resulted in confrontations between the community and local authorities. For example, the use of incense in the temple of Cham towers goes against the practice of religious beliefs, causing a significant conflict between the Cham community and local authorities. Despite the community's objections, the management board of the monument has not removed the incense cup from the temple but instead continues to leave it there for visitors from all over the world to burn incense.

Another issue that causes conflict is the building of the entrance to the tower from the east, which is reserved for the gods and is contrary to the traditional concept of direction in the traditional cultural philosophy of Cham. Despite community objections, the management continues to maintain its rationale for expansion to serve tourists and the Cham community, leading to continued conflicts. Furthermore, the scheduled performance of the Chăm people's New Year ceremony on the temple grounds to serve tourists has caused controversy. This ceremony is traditionally conducted within the community, not on the sacred temple grounds, due to differences in religious beliefs and practices. The Cham community protested against this as it distorts Chăm culture, leading to the heritage management board relocating the performance further away from the temple while still within the temple's premises. Lastly, the temple tower remains open every day, regardless of whether that day is considered bad in Cham culture, which is contrary to the traditional conception of time and taboo in the traditional cultural philosophy of Cham. This is a highly debated issue because the temple and tower must be open continuously to serve tourists. However, the Balamon Dignitaries Council also sent a letter requesting the closure before the important ceremony of the Cham community.

The temple serves as a symbol of the Cham people's unique cultural identity and religious beliefs. The cultural performances, while they may be a source of entertainment for tourists, can be seen as a disrespectful use of the sacred space. For the Cham people, the temple is not just a physical structure, but a living embodiment of their religious and cultural heritage. It is a place of worship, meditation, and reflection that connects them to their ancestors and the divine. The temple's sacredness is deeply ingrained in the Cham people's belief system and any action that could be perceived as a desecration of the temple is met with strong opposition.

The Cham community's concerns about the potential loss of the temple's sacredness reflect a broader issue faced by many indigenous and minority communities around the world. The encroachment of outside cultural and economic forces can threaten the integrity and survival of these communities' cultural traditions and beliefs. In the case of the Cham community, the use of the temple for cultural performances is just one example of a larger trend of cultural commodification that can result in the loss of spiritual and cultural identity.

The issues at hand underscore the disregard for the cultural traditions and beliefs of the Cham community in tourism development. The lack of consultation with the community leads to conflicts between them and the local authorities. Although the Cham community appreciates the Party and State's efforts in the conservation and promotion of their cultural heritage, they must be involved in the planning and execution of tourism activities within their sacred spaces. Failure to do so will result in ongoing conflicts, potentially leading to disillusionment with the work of the local management board.

Overall, the loss of sacredness at Cham temples is a multifaceted issue that requires a comprehensive approach to address. The consequences of not consulting or consulting improperly as it stands now are the formation of activities that do not align with Cham cultural traditions and create opposing reactions to the local government's tourism development efforts. Community consultation and participation in decision-making processes are crucial in upholding cultural values and beliefs, as well as promoting responsible and sustainable tourism practices. Failing to engage the community can have adverse effects on the genuineness and soundness of cultural heritage, as well as harm relationships between the local people, authorities, and tourists.

12.4.2 LOCAL GOVERNMENT'S RESPONSE TO CHAM COMMUNITY CONCERNS

Our research examined the government's response to the negative issues affecting the Cham community. Unfortunately, the responses we received were often general and not focused on specific issues, making it challenging to obtain comprehensive information on the solutions provided. While some authorities expressed positive thoughts toward finding solutions, the majority did not provide concrete steps. Meanwhile, Cham scholars have raised concerns about the lack of critical feedback providers in decision-making processes, resulting in ineffective participation by the Cham community. The Cham Heritage Management Unit proposed appointing specialists to manage the temple and prevent inappropriate offerings, as well as requiring worshippers to inform them in advance. Despite the Cham community's feedback during our interviews, many negative issues persist, such as tourists wearing revealing clothing. The government has proposed prohibiting it and providing suitable clothes for visitors at some heritage sites. Additionally, there is also no clear consensus on issues such as the new pathway from the east and the orientation method. The government has faced criticism for not listening to the voices of the Cham community regarding heritage management, and the Cham community strongly opposes incense use and other religious practices at their living temple. One of the drawbacks of the situation is the lack of effective management of the Cham temples. During our observations and discussions, many authorities avoided answering our questions directly and provided vague, unhelpful responses.

Overall, the involvement of critical feedback providers in decision-making and the establishment of clear regulations on state management of heritage units could help establish a good relationship between the Cham community and the authorities.

12.4.3 INADEQUATE INVOLVEMENT OF THE CHAM COMMUNITY IN DECISION-MAKING

The Cham community has had limited involvement in the decision-making process concerning their cultural heritage. Consequently, their opinions and perspectives have not been fully considered or included, and they have not had a chance to provide feedback on the preservation and management of their cultural heritage. Tourism development based on Cham cultural heritage resources, particularly the sacred sites that the community has maintained for spiritual and cultural purposes, requires respect for the traditional cultural essence of the heritage owners. Although the development of tourism is a necessary demand for the local government of Ninh Thuan province, which boasts unique cultural characteristics, it is essential to involve the Cham community in the planning and execution of tourism activities to ensure that their cultural heritage is preserved and honored. Therefore, Cham cultural resources in general and the sacred temples and shrines of the Cham community are also of concern. This concern is reflected in cultural heritage conservation policies and tourism product development. However, a prominent fact is that the local government does not seem to consult with the community in developing tourism in their sacred places and developing tourism products based on Cham's cultural and religious heritage. According to a member of the Ninh Thuan Cham Balamon Council:

"*We appreciate the Party and State's efforts in preserving and promoting Cham heritage. While we have seen significant investments in protective areas, temple reinforcement, and cultural heritage promotion, we are concerned about the lack of consultation and input from the local people in tourism development plans. Normally, decisions are made without our knowledge or input, leaving us to participate only for formality once everything has been completed. This approach leaves us with no power to effect change and does not consider the cultural traditions, religious beliefs, and needs of our community.*"

Another member of the Balamon Religious Council shared:

"*We raised concerns about practices like using incense and building eastward roads, but the local authorities only promised to consider our opinions and address them later. This issue has persisted for a decade, and our input seems to be only used for planning and feedback. We need a complete solution and a heard voice to avoid conflicts and discontent with the management board's work. Can we discuss the tourism development shortcomings at Cham sacred sites?*"

These statements reveal that the local government has already designed the plans for developing tourism products and activities in the Cham people's sacred temple, without seeking input from the Cham community, particularly the Balamon religious council. The community was only informed of the decisions after they were made. This highlights the limited and significant involvement of the Cham community, specifically the Balamon religious council, which lacks a substantial role in planning or executing tourism activities in the sacred spaces of the Cham people. Consequently, this observation suggests a lack of authentic community engagement in tourism planning, which could result in adverse effects on the Cham community's cultural heritage and sacred sites. Moreover, this may raise questions about the effectiveness of tourism governance and management, particularly in protecting cultural heritage and ensuring sustainable tourism development.

The findings reveal that the Cham Cultural Research Center plays a crucial role in supporting and advising the development of appropriate cultural tourism products, but has been largely overlooked in this capacity. The establishment of the Center reflects the state's interest in preserving and researching the Cham cultural heritage. The center is staffed with experts who are knowledgeable about the beliefs, religion, culture, and folk knowledge of the Cham people and are qualified to provide advice on heritage and cultural issues. Despite being designated as a consultant on Cham culture, the center's role has not received much attention, and the Cham community is rarely involved in decision-making processes related to cultural tourism development programs and projects in their sacred spaces. This represents a missed opportunity, as their involvement in tourism development consulting activities would undoubtedly bring valuable cultural insights to the table, enabling local authorities to make informed decisions that resonate with the aspirations and cultural traditions of the Cham community. In essence, the findings underscore the necessity for enhanced collaboration and communication among stakeholders engaged in the preservation and promotion of the value of local ethnic and cultural heritage. By involving the Cham Cultural Research Center more actively in decision-making processes, local officials can gain a better understanding of the needs and desires of the Cham population and respond accordingly, ensuring that cultural tourism development activities are culturally sensitive and respectful.

12.4.4 THE INTERVENTION OF THE LOCAL AUTHORITIES IN THE KATE FESTIVAL ON THE CHAM TEMPLES

The local government is heavily involved in the Cham temple festival, potentially having a significant impact on the event. The nature of this involvement may vary, which could include providing financial or logistical support for the festival, overseeing specific aspects of the event like security or transportation, or actively participating in the planning and organization of the festival. While such intervention may be well-intentioned and aimed at preserving and promoting the Cham culture, it is important for the local government to also respect the wishes and traditions of the Cham community and ensure that their involvement is appropriate and balanced.

The findings revealed that while there was effective collaboration between the Cham community and government authorities in organizing the Kate festival, there were also issues with government intervention in the festival that had negative impacts on the Cham community and their cultural practices. One major issue was the timing of the festival and the conduct of the ritual inside the temple. According to Cham zodiac hour, the ritual should have started earlier, but due to the time taken to introduce government representatives, the start time was delayed. This was seen by some Cham members as a violation of their beliefs and customs, as the ritual was meant to be performed at a specific time according to their traditional practices. Another issue was the way in which government officials were treated during the festival. While the Cham community welcomed their presence, some felt that the officials were given too much priority and that the festival had been turned into a state-run event, rather than a celebration of Cham culture. In particular, the order of events during the festival had been reversed, with the dance performed first to greet officials, followed by the traditional ritual. This was seen as a problem of reversal and a violation of the traditional order of events. My research findings found that Cham elders, scholars, and youths all expressed concerns about government intervention in the festival, calling for officials to respect their customs and traditions and not force them to change their practices to accommodate administrative procedures. Additionally, they stressed the significance of preserving the authenticity of the festival and upholding the traditional sequence of events.

Although many Chams express appreciation for the local government's significant interest in the culture of the Cham community, various perspectives shared by the Cham community indicate that this interest needs to be appropriately directed in order to realize the spirit and purpose of the government's cultural initiatives. The findings of this research show that the Cham community desires to receive fair and just treatment from the government in protecting and promoting their culture. According to a Cham researcher:

"*Provincial officials visiting the Cham community during the festival clearly show the interest of the State. However, right during the festival, the Cham people have to wait for each official to attend, which takes time and affects the sacred hour. Some officials are very enthusiastic, so they also arrive on time, and some are late and make people wait in the sun. Moreover, in my opinion, it is more reasonable to give gifts and perform administrative ceremonies before the opening of the traditional ceremony. I think this is also a suggestion for the local government to pay attention to, in order to respect the Cham tradition and simultaneously show the state's concern for the Cham people.*"

Despite these concerns, it is important to note that there was also effective collaboration between the Cham community and government authorities in organizing the festival. Heritage officers and the Cham Custom Board worked well with local authorities to ensure that the festival was organized effectively, and there was generally a positive attitude towards government officials who showed an interest in Cham culture and traditions.

Overall, the findings suggest that while collaboration between the Cham community and government authorities can be effective, it is important to be mindful of the potential negative impacts of government intervention on cultural practices and to work to ensure that these practices are respected and maintained.

12.4.5 POWER IMBALANCES IN MEETINGS AND PROBLEM, SOLVING SESSIONS

Meetings and problem-solving sessions are critical components of any organization's decision-making processes. However, one of the most significant challenges that organizations face is the presence of power imbalances within such gatherings. Power imbalances refer to the differences in power, influence, and authority between individuals or groups in a particular setting. According to research findings, there are still considerable challenges to achieving effective coordination between the government and the Cham community, despite the endeavors of the local authorities and functional departments to tackle the adverse effects of tourism development in the sacred sites of the Cham people. One major obstacle is the real power dynamic, where Cham people feel pressure to conform to the policies and projects of the Party and State, even if they disagree with them. This creates a difficult environment for open dialogue and honest feedback, as many Cham people are afraid to speak up for fear of consequences.

Furthermore, the research found that the interests and goals of the parties involved in tourism development often conflict, with economic benefits taking priority over cultural conservation and promotion. This makes it difficult to achieve a common consensus in solving numerous issues regarding tourism development in the sacred sites of the Cham people. Additionally, the research shows that most of the people invited to meetings on development policy or dealing with negative issues in tourism development are elderly Cham people with limited educational background. Cham intellectuals and experts with knowledge of customs and research have been invited to attend meetings in the past, but often do not feel comfortable speaking up about the shortcomings of heritage management and tourism, which may have contributed to their exclusion from future meetings.

It is clear that power imbalances are a significant issue in meetings and problem-solving sessions involving the Cham community and local authorities. This power imbalance can lead to a situation where individuals with more power and influence dominate the conversation, making it difficult for others to express their thoughts and views. As a result, some of the most valuable insights and contributions may go unheard, leading to suboptimal decision-making.

The issues highlighted in the research findings regarding the coordination between the Cham community and the local government and related departments are not unique to this context. In fact, previous studies have also identified power differentials and obstacles to effective communication and collaboration between communities and government agencies in the context of heritage conservation and tourism development. One study by Zhang & Wang (2017) on cultural heritage management in China found that local communities often feel marginalized and powerless in the decision-making process, which can lead to conflicts with government agencies. This is similar to what was found in the Cham community, where many interviewees expressed feeling pressure and fear when working with the government due to the power dynamics involved.

In terms of the focus on economic benefits over cultural conservation and promotion, this issue has also been identified in previous research on tourism development in heritage sites. For example, Kim & Jamal (2007) found that there is often a conflict between the economic benefits of tourism and the conservation of cultural heritage sites. This can lead to a focus on increasing visitor numbers and revenue, rather than on preserving and promoting the cultural significance of the site. In the case of the Cham community, the fact that many programs and projects have been implemented in the sacred heritage area for tourism development suggests that economic benefits may be taking precedence over cultural conservation and promotion. This is a complex issue that necessitates a balance between economic growth and heritage conservation, which is a challenge faced by many heritage sites around the world. Moreover, the pressure on Cham intellectuals and dignitaries not to speak up against policy decisions is a concerning issue. This is a violation of the freedom of speech and the right to express their opinions.

During my research, it was evident that individuals with a gender education background were primarily invited to meetings regarding development policy or negative issues in tourism development. The majority of these individuals were elderly people, dignitaries, and farmers who were passionate about preserving Cham culture. However, due to their lack of confidence, they struggle to voice their views during these meetings. According to two Cham research experts knowledgeable about customs and research, they were invited to multiple meetings in the past to resolve outstanding issues in tourism development at Cham temples. However, they often encountered resistance when they spoke openly about the shortcomings of heritage management and tourism. This response made them uncomfortable, and consequently, many Cham intellectuals were not invited to attend subsequent meetings. Therefore, it is essential for authorities to create a safe and inclusive environment where all individuals can voice their opinions and concerns, regardless of their age or gender. This will not only benefit the preservation of cultural heritage but will also improve the effectiveness of tourism development policies.

Overall, the issues identified in the research findings regarding the coordination between the Cham community and the local government and related departments are not unique to this context and have been documented in previous research on heritage conservation and tourism development. However, these issues are complex and require careful consideration and collaboration to achieve a balance between economic development and cultural preservation. Additionally, the issue of pressure on intellectuals and dignitaries not to speak up against policy decisions is concerning and requires attention to ensure freedom of speech and the right to express opinions.

12.5 DISCUSSION AND IMPLICATIONS

Firstly, the limited involvement of the Cham community in decision-making regarding their cultural heritage is a common issue in indigenous community tourism. This finding is similar to research on indigenous communities in Mexico and Peru, where community members conveyed a sense of being left out of the planning and administration of tourism activities in their area, leading to the cultural commodification of their traditions and beliefs for profit (Coronado, 2014). Conversely, research has emphasized the significance of involving indigenous communities in tourism planning and management, as community participation in tourism development has resulted in favorable outcomes for both the community and visitors in certain cases (Mbaiwa, 2011). The Cham community's limited role in tourism product development underscores the significance of strengthening collaboration among stakeholders involved in tourism planning and management. By fostering greater participation from all relevant parties, tourism activities can be designed and executed in a manner that respects and preserves the Cham community's cultural heritage while simultaneously promoting sustainable and responsible tourism practices. Tosun (2006) stressed the significance of involving the local community in the decision-making process of tourism development to guarantee that the development caters to the community's needs and desires and that the tourism activities are sustainable in the long run. Additionally, previous research has identified a problem with the deficient coordination and communication among stakeholders in the management of living heritage sites (Arkarapotiwong, 2015; Pearson & Sullivan, 1995). Several factors can be attributed to this lack of coordination, such as the historical disregard for the role of the core community in heritage management and the different priorities and interests of stakeholders (Poulios, 2014). Therefore, scholars have called for a more collaborative approach to heritage management, where the core community is placed at the center of the process, and stakeholders work together to ensure that the community's cultural values and well-being are respected (Sarah & Wijesuriya, 2014). This approach can help to ensure that community concerns and needs are addressed and can prevent conflicts from escalating.

Secondly, the issue of government intervention in cultural festivals is not unique to the Cham community and has been discussed in previous research. A study on the impact of government involvement in the Samba festival in Brazil found that the festival had become overly commercialized and focused on tourism, with the cultural elements of the event being overshadowed by the government's emphasis on economic development (Tilley, 1997; Zhang, 2021). In a similar vein, a study on the government's involvement in two tourist villages in China's southwestern province of Guizhou revealed that, despite the government's crucial role in organizing the event, certain community members felt that the festival had become excessively commercialized, leading to the loss of its traditional significance (Oakes, 1997). The outcomes of this research conform to earlier studies that suggest government intervention can have adverse effects on the cultural heritage and identity of the community. For the Cham community, there were apprehensions about the negative influence of government intervention on the festival, particularly with regard to the timing of the ceremony and the sequence of events. These findings underscore the significance of acknowledging and honoring the cultural values and beliefs of the host community to prevent any negative repercussions on their cultural customs and traditions.

Furthermore, the concept of cultural sensitivity is essential for successful collaborations between stakeholders in ethnic or indigenous heritage and tourism development. It involves acknowledging the cultural heritage and identity of the host community, including their cultural perspectives in tourism development, and promoting cultural preservation and conservation (Pearce, Moscardo, & Ross, 1996). Failure to recognize the cultural sensitivity of the host community can result in negative impacts on the community's cultural heritage and identity, as well as their economic and social well-being (Murphy, 1988; Tuyen Dai Quang, 2022). In short, collaboration between the community and government authorities is crucial in ensuring that cultural festivals are organized effectively and that the traditions and beliefs of the community are respected and maintained. The research findings offer significant insights into the possible unfavorable effects of government intervention on cultural customs and traditions. By taking a collaborative approach that prioritizes cultural sensitivity, it is possible to promote tourism development that benefits both the host community and tourists.

Thirdly, meetings and problem-solving sessions are crucial components of an organization's decision-making process, but power imbalances can create significant challenges within such gatherings. Research has shown that power imbalances exist in meetings involving the Cham community and local authorities, hindering effective coordination in addressing issues related to tourism development and cultural preservation. Many Cham people feel pressure to conform to the policies and projects of the Party and State, leading to a difficult environment for open dialogue and honest feedback. The existing dynamic presents hindrances in attaining a mutual agreement in resolving numerous concerns associated with tourism development in the sacred spaces of the Cham community. Additionally, the research shows that the interests and goals of the parties involved in tourism development often conflict, with economic benefits taking priority over cultural conservation and promotion. Similar findings have been reported in previous studies on the management of cultural heritage sites and tourism development. For instance, Yang & Wall (2009) found that local communities often feel marginalized and powerless in the decision-making process, leading to conflicts with government agencies. Moreover, the focus on economic benefits over cultural conservation and promotion has also been identified in previous research on tourism development in heritage sites (Coronado, 2014; Johnston, 2006; Tuyen Dai Quang, 2022). This can lead to a focus on increasing visitor numbers and revenue, rather than preserving and promoting the cultural significance of the site. The challenge of balancing economic development and cultural preservation is a common issue faced by numerous heritage sites across the globe.

Another issue highlighted in the research is the exclusion of Cham intellectuals and experts from meetings related to tourism development and cultural heritage conservation. Cham intellectuals and experts with knowledge of customs and research have been invited to attend meetings in the past, but they often do not feel comfortable speaking up about the shortcomings of heritage management and tourism. This may have contributed to their exclusion from future meetings. This issue is not unique to the Cham community, and similar challenges have been documented in other contexts as well (Hitchcock, King, & Parnwell, 2009). Furthermore, individuals with more power and influence can dominate the conversation during meetings, making it difficult for others to express their opinions and ideas. This can result in valuable insights and contributions going unheard, leading to suboptimal decision-making. Power relations among stakeholders in the tourism industry, particularly in ethnic or indigenous heritage contexts, have been identified as a significant challenge for successful collaboration (Carr et al., 2016a; Crawford, 2001). Power imbalances among stakeholders can result in the marginalization of the host community and a lack of participation in decision-making (Cornelisse, 2020; Dong & Nguyen, 2022; Saito & Ruhanen, 2017). The economic and political influence of stakeholders, particularly prominent tourism companies and the government, often affects power dynamics (Nunkoo, 2017; Salazar, 2012). However, power relations can also be influenced by cultural and social factors, such as the historical relationship between the host community and the tourism industry and the recognition of the host community's cultural values and beliefs (Cornelisse, 2020; Dong & Nguyen, 2022; Saito & Ruhanen, 2017).

In short, power imbalances in meetings and decision-making processes can hinder effective coordination and lead to suboptimal decision-making. The coordination challenges between the Cham ethnic group and the local authorities and relevant departments that the research findings have brought to light are not exclusive to this context. Similar difficulties have been noted in other settings concerning tourism development and the conservation of cultural heritage. Achieving a balance between economic development and cultural preservation is a challenge faced by many heritage sites around the world.

Finally, the roles of different stakeholders in collaborations between stakeholders in ethnic or indigenous heritage and tourism development are critical to their success. Different stakeholders, such as the local people, government, tourism businesses, non-governmental organizations, and tourists, have different interests, values, and perspectives that must be recognized and addressed (Davis, Huang, & Liu, 2010; Jamal & Stronza, 2009). The host community plays a critical role in the management and development of tourism in their communities, and their participation in decision-making is essential for the sustainability of collaborations (Jamal & Stronza, 2009; Mak et al., 2017; Page, Moscardo, & Ross, 1996). However, the inclusion of the host community in decision-making processes is often limited, and power imbalances can arise between various stakeholders (Cornelisse, 2020; Dong & Nguyen, 2022; Saito & Ruhanen, 2017). The tourism industry often focuses on economic benefits, which can lead to the neglect of the cultural and social impacts on host communities (Mason, 2015; Timothy & Nyaupane, 2009). This situation can give rise to conflicts between stakeholders and lead to unfavorable consequences for both the local people and the tourism industry (Quang et al., 2022; Yang & Wall, 2009). To overcome these challenges, it is essential for all stakeholders to work together and engage in effective communication and collaboration. Successful collaboration necessitates comprehending the cultural and social context of the host community, as well as their requirements and anticipations concerning tourism development and heritage management (Carr et al., 2016b; Jamal & Stronza, 2009). Moreover, the participation of all stakeholders in decision-making processes and acknowledging the contributions of each stakeholder are crucial to guarantee the durability of collaborations (Dabphet, 2016). This may involve offering training and capacity-building opportunities for the host community to participate efficiently in decision-making and to manage and promote their cultural heritage in a sustainable manner (Mak et al., 2017).

In summary, collaboration among diverse stakeholders is vital for the sustainable development of ethnic or indigenous heritage and tourism. Inclusion of the host community in decision-making processes, acknowledging the contributions of all stakeholders, and effective communication and collaboration are crucial for successful collaborations. Additionally, it is crucial to consider the cultural and social context of the host community and to balance economic growth and cultural conservation.

12.6 CONCLUSION

The findings reveal that there are diverse perceptions and attitudes toward tourism development at Cham temples among different stakeholders. While most stakeholders support tourism development, the Cham community has concerns about the lack of controlled access, funding for ceremonies, and protection and preservation of their sacred site. The government has made efforts to protect and preserve the Cham temples, but there have been negative incidents and issues, such as inappropriate behavior, misuse of the temple, and misinterpretations of Cham culture. The Cham community acknowledges the role of government but also feels marginalized in planning processes. Additionally, this study has indicated that power disparities, conflicting interests and objectives, and the exclusion of Cham intellectuals and experts from meetings concerning tourism development and cultural heritage conservation are significant impediments to attaining effective coordination and decision-making within the Cham people. The findings highlight the need to address power imbalances and promote inclusive and participatory decision-making processes that involve all stakeholders, including the Cham community, local government and related departments, and tourism businesses. The task of reconciling economic development with cultural preservation is a widespread issue faced by numerous heritage sites globally.

This research sheds light on the difficulties that ethnic minority communities face concerning tourism development and cultural heritage conservation. The study has enhanced our comprehension of power dynamics among stakeholders in the tourism industry, particularly in ethnic or indigenous heritage settings, and how they can affect the efficacy of collaborations. The outcomes can guide policymakers and stakeholders in the tourism industry in promoting sustainable and inclusive tourism development in ethnic minority communities. Additionally, the study has emphasized the necessity for further research on the experiences of ethnic minority communities in the tourism industry, particularly concerning power imbalances and participatory decision-making processes.

KEYWORDS

- **cham people**
- **collaborations**
- **compound annual growth rate**
- **decision-making**
- **heritage conservation**
- **living heritage approach**
- **stakeholders**
- **sustainable tourism development**
- **tourism development**

REFERENCES

Aas, C., Ladkin, A., & Fletcher, J. (2005). Stakeholder collaboration and heritage management. *Annals of Tourism Research, 32*(1), 28–48. https://doi.org/10.1016/j.annals.2004.04.005.

Bin Abdul Hamid, M. E. (2013). *Nager Cam & the priests of prowess: A history of resilience* (Doctoral dissertation, University of Hawaii at Manoa).

Arkarapotiwong, P. (2015). *The Investigation of Living Heritage Attributes in Living Heritage sites: Case Study: Nan, Thailand and Luang Prabang, Lao PDR* (Doctoral dissertation).

Balen, K. Van, & Vandesande, A. (2015). *Community Involvement in Heritage Management* (Reflections on Cultural Heritage Theories and Practices Vol. 10). Garant Uitgevers nv.

Bernard, H. R. (2011). *Research Methods in Anthropology* (5th ed.). AltaMira Press.

Bien, P. X., An, P., & Dop, P. Van. (1991). *The Cham Culture [Văn hoá Chăm].* Ho Chi Minh: Social Sciences Publishing House.

Bornhorst, T., Brent Ritchie, J. R., & Sheehan, L. (2010). Determinants of tourism success for DMOs & destinations: An empirical examination of stakeholders' perspectives. *Tourism Management, 31*(5), 572–589. https://doi.org/10.1016/j.tourman.2009.06.008.

Borrini-Feyerabend, G., Farvar, M. T., Nguinguiri, J. C., & Ndangang, V. A. (2007). *Comanagement of Natural Resources: Organising, Negotiating and Learning-by-doing.* GTZ and IUCN, Kasparek Verlag.

Bradford, M., & Lee, E. (2004). *Tourism and cultural heritage in Southeast Asia.*

Bramwell, B. (2010). Participative planning and governance for sustainable tourism. *Tourism Recreation Research, 35*(3), 239–249. https://doi.org/10.1080/02508281.2010.11081640.

Bruner, E. M. (2005). *Culture on tour: Ethnographies of travel.* The University of Chicago Press.

Buckley, R. (2012). Sustainable tourism: Research and reality. *Annals of Tourism Research, 39*(2), 528–546. https://doi.org/10.1016/j.annals.2012.02.003.

Bui, H. T., Jones, T. E., Weaver, D. B., & Le, A. (2020). The adaptive resilience of living cultural heritage in a tourism destination. *Journal of Sustainable Tourism, 28*(7), 1022–1040. https://doi.org/10.1080/09669582.2020.1717503.

Byrd, E. T. (2007). Stakeholders in sustainable tourism development and their roles: Applying stakeholder theory to sustainable tourism development. *Tourism Review, 62*(2), 6–13. https://doi.org/10.1108/16605370780000309.

Carr, A., Ruhanen, L., & Whitford, M. (2016a). Indigenous peoples and tourism: The challenges and opportunities for sustainable tourism. *Journal of Sustainable Tourism, 24*(8–9), 1067–1079. https://doi.org/10.1080/09669582.2016.1206112.

Carr, A., Ruhanen, L., & Whitford, M. (2016b). Indigenous peoples and tourism: The challenges and opportunities for sustainable tourism. *Journal of Sustainable Tourism, 24*(8–9), 1067–1079. https://doi.org/10.1080/09669582.2016.1206112.

Cornelisse, M. (2020). Peru case study: Power relations in community-based tourism. *Journal of Tourism and Cultural Change, 18*(4), 437–454. https://doi.org/10.1080/14766825.2019.1640707.

Coronado, G. (2014). Selling culture? Between commoditisation and cultural control in Indigenous alternative tourism. *PASOS Revista de Turismo y Patrimonio Cultural, 12*(1), 11–28. https://doi.org/10.25145/j.pasos.2014.12.002.

Crawford, A. (2001). Successful tourism at heritage places. *Info, 50–53.* https://doi.org/AHC20.

Creswell, J. W. (2014). *Research Design: Qualitative, Quantitative and Mixed Methods Approaches* (4th ed.). Thousand Oaks, CA: Sage.

Dabphet, S. (2013). *The Key Stakeholders in The Implementation of Sustainable Tourism Development in Two Rural Towns of Thailand.* Department of Tourism, Faulty of Business. Economics and Communications, Naresuan University, Thailand.

Davis, P., Huang, H.-Y., & Liu, W.-C. (2010). Heritage, local communities, and the safeguarding of "Spirit of Place" in Taiwan. *Museum and Society, 8*(2), 80–89. Retrieved from http://openarchive.icomos.org/84/1/77-LtNt-2012_.pdf.

Dharma, P. (2013). *Vương quốc Champa: lịch sử 33 năm cuối cùng*. International Office of Champa, San Jose, CA.

Dolezal, C., & Novelli, M. (2022). Power in community-based tourism: Empowerment and partnership in Bali. *Journal of Sustainable Tourism, 30*(10), 2352–2370. https://doi.org/10.1080/09669582.2020.1838527.

Dong, X. D., & Nguyen, T. Q. T. (2023). Power, community involvement, and sustainability of tourism destinations. *Tourist Studies, 23*(1), 62–79. https://doi.org/10.1177/14687976221144335.

Dwyer, L., Gill, A., & Seetaram, N. (Eds.). (2012). *Handbook of research methods in tourism: Quantitative and qualitative approaches.* Edward Elgar Publishing.

Garrod, B., & Fyall, A. (2000). Managing heritage tourism. *Annals of Tourism Research, 27*(3), 682–708. https://doi.org/10.1016/S0160–7383(99)00094–8.

General Statistics Office. (2019). *Completed Result of the 2019 Viet Nam Population and Housing Census.* Statistics Publishing House.

Gurcayir Teke, S. (2018). Living heritages and fixed traditions in the lists of the convention for the safeguarding of intangible cultural heritage. *Milli Folklor,* (120), 19–31.

Heslinga, J., Groote, P., & Vanclay, F. (2021). Strengthening governance processes to improve benefit-sharing from tourism in protected areas by using stakeholder analysis. In *Living on the Edge* (pp. 69–83). Routledge. https://doi.org/10.1080/09669582.2017.1408635.

Hitchcock, M., King, V. T., & Parnwell, M. (Eds.). (2009). *Tourism in Southeast Asia: Challenges and new directions.* Nias Press.

Jamal, T., & Stronza, A. (2009). Collaboration theory and tourism practice in protected areas: Stakeholders, structuring and sustainability. *Journal of Sustainable Tourism, 17*(2), 169–189. https://doi.org/10.1080/09669580802495741.

Johnston, A. M. (2006). *Is the Sacred for Sale: Tourism and Indigenous Peoples.* Routledge.

Kim, H., & Jamal, T. (2007). Touristic quest for existential authenticity. *Annals of Tourism Research, 34*(1), 181–201. https://doi.org/10.1016/j.annals.2006.07.009.

Langfield, M., Logan, W., & Craith, M. N. (2010). *Cultural diversity, heritage and human rights.* Abingdon and New York: Routledge. https://doi.org/10.4324/9780203863015.

Lenzerini, F. (2011). Intangible cultural heritage: The living culture of peoples. *European Journal of International Law, 22*(1), 101–120. https://doi.org/10.1093/ejil/chr006.

Li, Y., & Hinch, T. (1997). Ethnic tourism attractions and their prospect for sustainable development at two sites in China and Canada. *Asia Pacific Journal of Tourism Research, 2*(1), 1094–1665. https://doi.org/10.1080/10941669808721981.

Lu, J., & Nepal, S. K. (2009). Sustainable tourism research: An analysis of papers published in the Journal of Sustainable Tourism. *Journal of Sustainable Tourism, 17*(1), 5–16. https://doi.org/10.1080/09669580802582480.

Mak, B. K., Cheung, L. T., & Hui, D. L. (2017). Community participation in the decision-making process for sustainable tourism development in rural areas of Hong Kong, China. *Sustainability, 9*(10), 1695. https://doi.org/10.3390/su9101695.

Mason, P. (2020). *Tourism impacts, planning and management.* Routledge. https://doi.org/10.4324/9781315781068.

Mbaiwa, J. (2011). Cultural commodification and tourism: The Goo-Moremi community, Central Botswana. *Tijdschrift Voor Economische En Sociale Geografie, 102*(3), 290–301. https://doi.org/10.1111/j.1467–9663.2011.00664.x.

Miura, K. (2005). Conservation of a 'living heritage site' A contradiction in terms? A case study of Angkor World Heritage Site. *Conservation and Management of Archaeological Sites, 7*(1), 3–18. https://doi.org/10.1179/135050305793137602.

Murphy, P. E. (1988). Community driven tourism planning. *Tourism Management, 9*(2), 96–104. https://doi.org/10.1016/0261–5177(88)90019–2.

Nunkoo, R. (2017). Governance and sustainable tourism: What is the role of trust, power and social capital? *Journal of Destination Marketing and Management, 6*(4), 277–285. https://doi.org/10.1016/j.jdmm.2017.10.003.

Oakes, T. S. (1997). Ethnic tourism in rural Guizhou: Sense of place and the commerce of authenticity. In M. Picard & R. E. Wood (Eds.), *Tourism, ethnicity, and the state in Asian and Pacific societies* (pp. 35–70). University of Hawai'i Press.

Page, P., Moscardo, G., & Ross, G. (1996). *Tourism community relationships.* Emerald Publishing Limited.

Patton, M. Q. (2002). *Qualitative research & evaluation methods.* Sage.

Pearce, P. L., Moscardo, G., & Ross, G. F. (1996). *Tourism community relationships.* Oxford: Pergamon.

Pearson, M., & Sullivan, S. (2013). *Looking after heritage places: The basics of heritage planning for managers, landowners and administrators.* Melbourne Univ. Publishing.

Poulios, I. (2014). *Past in the Present.* Ubiquity Press. https://doi.org/10.5334/bak.

Quang, Tuyen D., Noseworthy, W. B., & Paulson, D. (2022). Rising tensions: Heritage-tourism development and the commodification of "Authentic" culture among the Cham community of Vietnam. *Cogent Social Sciences, 8*(1), 2116161. https://doi.org/10.1080/23311886.2022.2116161.

Quang, T. D. (2022). Is Living Heritage Ignored? Revisiting Heritage Conservation at Cham Living-Heritage Sites in Vietnam. *Heritage & Society, 15*(1), 46–74. https://doi.org/10.1080/2159032X.2022.2126234.

Renault, M. (2008). Tell me about.. living heritage. UNESCO: United Nations Educational, Scientific and Cultural Organisation. France. Retrieved from https://policycommons.net/artifacts/8880879/tell-me-about/9732102/ on 25 Jun 2024. CID: 20.500.12592/wpzgncn.

Saito, H., & Ruhanen, L. (2017). Power in tourism stakeholder collaborations: Power types and power holders. *Journal of Hospitality and Tourism Management, 31*, 189–196. https://doi.org/10.1016/j.jhtm.2017.01.001.

Sakaya. (2003). *The festivals of the Cham people [Lễ hội của người Chăm].* Culture of Vietnamese Ethnic Groups Publishing House.

Salazar, N. B. (2012). Community-based cultural tourism: Issues, threats and opportunities. *Journal of Sustainable Tourism, 20*(1), 9–22. https://doi.org/10.1080/09669582.2011.596279.

Sarah, C., & Wijesuriya, G. (2014). People-centred approaches to the conservation of cultural heritage: Living heritage.

Sharpley, R. (2000). Tourism and sustainable development: Exploring the theoretical divide. *Journal of Sustainable Tourism, 8*(1), 1–19. https://doi.org/10.1080/09669580008667346.

Silva, K. D., & Chapagain, N. K. (Eds.). (2013). *Asian heritage management: Contexts, concerns, and prospects.* Routledge.

Tilley, C. (1997). Performing culture in the global village. *Critique of Anthropology, 17*(1), 67–89. https://doi.org/10.1177/0308275X9701700105.

Timothy, D. J., & Nyaupane, G. P. (Eds.). (2009). *Cultural heritage and tourism in the developing world.* New York: Routledge.

Tosun, C. (2006). Expected nature of community participation in tourism development. *Tourism Management, 27*(3), 493–504. https://doi.org/10.1016/j.tourman.2004.12.004.

Tuyen, Q. D. (2019). Phát triển bền vững các sản phẩm du lịch văn hoá Chăm ở tỉnh Ninh Thuận: Tiếp cận các bên liên quan. *Central Vietnamese Review of Social Sciences, 6*, 62–76.

UNWTO. (2015). *Tourism and the Sustainable Development Goals.* UNWTO.

Walter, N. (2020). Narrative approach to living heritage. *Protection of Cultural Heritage, (10), 126–138.* https://doi.org/10.35784/odk.2443.

Weise, K. (2013, November). Revisiting Kathmandu: Safeguarding living urban heritage. In *Proceeding of an International Symposium, Kathmandu Valley* (pp. 25–29).

Wijesuriya, G. (2007). Conserving living Taonga: The concept of continuity. In D. Sully (Ed.), *Decolonizing conservation: Caring for Maori meeting houses outside New Zealand (Critical Cultural Heritage Series)* (pp. 59–69). London: Left Coast Press Inc.

Wijesuriya, G. (2014). Introducing people-centred approach to conservation and management of Hani Rice Terraces. In *International Workshop on the Sustainable Development of Honghe Hani Terraces* (pp. 23–34). Mengzi: ICOMOS China.

Wijesuriya, G. (2015). *Living heritage: A summary.* Rome: ICCROM.

Yanes, A., Zielinski, S., Diaz Cano, M., & Kim, S. (2019). Community-based tourism in developing countries: A framework for policy evaluation. *Sustainability, 11*(9), 2506. https://doi.org/10.3390/su11092506.

Yang, L., & Wall, G. (2009). Minorities and tourism: Community perspectives from Yunnan, China. *Journal of Tourism and Cultural Change, 7*(2), 77–98. https://doi.org/10.1080/14766820902849971.

Zhang, Y. (2021). Transnational religious tourism in modern China and the transformation of the cult of Mazu. *Religions, 12*(3), 221. https://doi.org/10.3390/rel12030221.

CHAPTER 13

PRESERVING INTANGIBLE CULTURAL HERITAGE OF KHASI INDIGENOUS COMMUNITY OF MEGHALAYA THROUGH NONGKREM DANCE FESTIVAL: AN EVENT TOURISM DESTINATION PERSPECTIVE

IADONLANG TYNSONG[1] and
SHARAD KUMAR KULSHRESHTHA[2]

[1]*Faculty, Department of Tourism and Travel Management, Martin Luther Christian University, Shillong, Meghalaya, India*

[2]*Assistant Professor, Department of Tourism and Hotel Management, North-Eastern Hill University, Shillong, Meghalaya, India*

ABSTRACT

The Nongkrem Festival is an important intangible cultural heritage of the Khasi people, and it is crucial to preserve its sacredness and traditional values while promoting it as a tourist destination. Event tourism can play a significant role in achieving this goal by showcasing the festival to a wider audience without altering its cultural significance. The main aim of the study is to identify the challenges faced by the Khasi community in preserving their intangible cultural heritage in the face of modernization and globalization, to explore the potential of event tourism in promoting the Nongkrem Festival as a tourist destination while ensuring the preservation of its cultural values and traditions, and to develop guidelines and recommendations for promoting

Dynamics of Heritage Management and Tourism Development: Trends, Practices, and Approaches.
Surabhi Srivastava & Ruchika Kulshrestha (Eds.)

the Nongkrem Festival as an event tourism destination that balances the need for economic development with the preservation of cultural heritage. A qualitative design was used for conducting the study, and a deductive approach was used to draw the conclusions. The major finding of the study is that the Khasi community faces challenges in preserving their cultural heritage, such as the impact of modernization and globalization, the lack of resources, and the need to balance tradition with economic development. In conclusion, the Nongkrem Festival is an important cultural heritage of the Khasi people, and event tourism can play a significant role in promoting it as a destination without compromising its sacredness. By working together, the local authorities, tourism officials, and local communities can ensure that the festival is preserved for future generations while also promoting the growth and development of the area.

13.1 INTRODUCTION

The Khasi indigenous community of Meghalaya has a rich cultural heritage, including their traditional dance forms, which are an important aspect of their social and religious life. One of the most popular dance forms among the Khasi people is the Nongkrem dance, which is performed during the annual Nongkrem dance festival (Basaiamoit & Khyriemmujat, 2016). The festival is an important event in the Khasi calendar, and it is a significant tourist attraction that draws visitors from around the world. This literature review examines the significance of the Nongkrem dance festival in preserving the intangible cultural heritage of the Khasi community and its potential as an event tourism destination.

The preservation of intangible cultural heritage has become an important issue in recent years, as globalization and modernization threaten to erode traditional cultural practices. According to UNESCO (2014), intangible cultural heritage includes traditions, practices, and expressions passed down from generation to generation and rooted in a community's cultural identity. The Nongkrem dance festival is an important way for the Khasi community to preserve their intangible cultural heritage, as it is an opportunity to showcase their traditional dance forms and other cultural practices.

Event tourism has become an increasingly popular form of tourism in recent years, and festivals and other cultural events are important draws for tourists (Getz, 2008). The Nongkrem dance festival has the potential to be a significant event tourism destination, as it offers visitors a unique cultural experience that cannot be found anywhere else in the world. The festival is a celebration of the Khasi community's cultural identity, and it provides an opportunity for visitors to learn about and experience the community's traditional practices and customs.

Several studies have explored the potential of festivals and cultural events as tourism destinations. For example, according to a study by Getz and Page (2016), festivals can help to promote local economic development and create employment opportunities. Another study states that festivals can help to promote social cohesion and community development (Mair & Duffy, 2014). The Nongkrem dance festival can potentially contribute to economic and social development in the Khasi community, as it can create employment opportunities for local people and promote social cohesion by bringing the community together.

13.2 LITERATURE REVIEW

Preserving intangible cultural heritage has become an important issue in cultural policy and heritage management, particularly in the last few decades. The literature on this topic covers a wide range of areas, including the definition and significance of intangible cultural heritage, the challenges of preserving it, and the various strategies and approaches that have been developed to safeguard it.

One of the key themes in the literature is the need to recognize the value of intangible cultural heritage and its role in shaping cultural identity and social cohesion. As argued by scholars such as Smith & Akagawa (2009), intangible cultural heritage represents the living cultural heritage of communities, and preserving it is crucial for maintaining cultural diversity and promoting intercultural dialogue. This recognition has led to the development of international legal frameworks, such as the UNESCO (2014) guidelines for the identification, documentation, and protection of intangible cultural heritage.

However, preserving intangible cultural heritage also poses significant challenges, particularly in the face of globalization, urbanization, and social change. These challenges include the loss of traditional knowledge and skills, the erosion of cultural practices and values, and the lack of resources and expertise for documentation and conservation. Scholars such as Hafstein (2012) have highlighted the need to develop new strategies and approaches that take into account the dynamic nature of intangible cultural heritage and the changing social contexts in which it is embedded.

13.2.1 NONGKREM DANCE AND ITS ROLE IN PRESERVING THE CULTURAL IDENTITY

The Khasi indigenous community of Meghalaya, India, is known for its rich cultural heritage and traditions, including the Nongkrem Dance Festival, Shadsuk Mynsiem, etc. The Nongkrem Dance Festival is an annual event that takes place in the village of Smit, Meghalaya and is celebrated by the Khasi people. The festival is a symbol of the community's rich cultural heritage and is characterized by traditional music, dance, and rituals. Tynsong (2019) has highlighted the significance of the festival in promoting the cultural identity of the Khasi community, particularly in the face of globalization and modernization. As the festival is considered an important symbol of the cultural identity of the Khasi community and provides a platform for the community to address the challenges of preserving their cultural heritage in the face of social and economic change.

Most of the literature on the Nongkrem Dance Festival discusses the various cultural practices and rituals that are associated with the festival. These include the traditional dress worn by participants, the offerings made to the gods, and the various dances that are performed. Lyngdoh (2018) has emphasized the importance of these practices in preserving the cultural heritage of the Khasi community and promoting intercultural dialog. They also highlight the challenges of preserving traditional cultural practices in the face of social and economic change. These challenges include the loss of traditional knowledge and skills, the erosion of cultural values, and the impact of globalization on local cultural practices. Sawian (2012) has argued that the festival provides a platform for the Khasi community to address these challenges and to promote the preservation and transmission of their cultural heritage to future generations. In response to these challenges, a range of preservation strategies and approaches have been developed. These include community-based approaches that involve local communities in the documentation and safeguarding of their intangible cultural heritage, as well as participatory approaches that involve collaboration between communities and cultural institutions. Liu (2022) has also emphasized the importance of digital technologies in preserving intangible cultural heritage, particularly in terms of facilitating access to and dissemination of cultural knowledge and practices.

However, there are also challenges associated with using festivals and cultural events as tourism destinations. For example, there may be issues around cultural authenticity, as events may become commercialized and lose their cultural significance. In addition, there may be issues around sustainability, as events can have a negative impact on the environment and local communities if they are not managed properly. It is important that the Nongkrem dance festival is managed in a way that preserves its cultural authenticity and ensures that it has a positive impact on the Khasi community and the environment.

In conclusion, the Nongkrem dance festival is an important way for the Khasi community to preserve their intangible cultural heritage and has the potential to be a significant event tourism destination. However, it is important that the festival is managed in a way that preserves its cultural authenticity and ensures that it has a positive impact on the Khasi community and the environment. Further research is needed to explore the potential of the Nongkrem Dance Festival as an event tourism destination and to identify strategies for managing the festival in a sustainable way.

13.3 RESEARCH GAP

To examine the Nongkrem festival and its significance as a cultural heritage event, it is pointed out that the festival is recognized as an important cultural event, but there is a lack of research on the festival's impact from a destination perspective. This highlights a gap in research on the tourism potential of the festival as a cultural destination (Dychkovskyy & Ivanov, 2020). Similarly, Dutta (2019) noted that while tourism has the potential to promote the preservation of cultural heritage, there is a lack of research on the role of intangible cultural heritage in sustainable tourism development in Northeast India. Additionally, Mon, & Thanikkad, and Jaseela (2021) note that there is a need for research on the relationship between intangible cultural heritage and tourism development in the region. While the Nongkrem festival is recognized as an important cultural heritage event, overall, these studies highlight a research gap in the potential for the Nongkrem festival to promote sustainable tourism development and the preservation of intangible cultural heritage in Northeast India. There is a need for further research on the festival's impact from a destination perspective and its potential to promote sustainable tourism development and the preservation of intangible cultural heritage in the region. Such research can help promote the cultural authenticity of the festival and ensure its long-term sustainability as a cultural heritage event.

While there has been some research on the Nongkrem dance festival and its potential as an event tourism destination, there are still several research gaps that need to be addressed. Some potential research gaps and questions for future research on this topic include:

Cultural authenticity: Sawain (2012) highlights the importance of the Nongkrem festival in Khasi culture and tradition. They note that the festival reflects the Khasi community's close relationship with the land and its resources and is deeply connected to their agricultural practices. The authors also describe the different rituals and ceremonies associated with the festival and how they are performed by the community. Similarly, Tynsong (2019) discusses the cultural significance of the Nongkrem festival in Khasi society. They note that the festival has been celebrated for centuries and has been passed down through generations, reflecting the community's traditional beliefs and practices. The authors also describe the different elements of the festival, including the sacrificial offering and the traditional dance, and how they are performed. While the festival has remained an important part of Khasi culture, it has also evolved to adapt to changing social and economic conditions. The authors also describe the role of the Syiem and the community in the performance of the festival and how it reflects the Khasi community's sense of identity and belonging (Dutta & Ghosh, 2016).

Sinha & Murdia (2018) examine the economic impacts of tourism. The authors note that the festival has the potential to attract a significant number of tourists to the region, generating economic benefits for the local community. They also highlight the importance of sustainable tourism practices to ensure that the economic benefits are distributed equitably and that the festival's cultural authenticity is preserved. Nayak & Mishra (2013) note that while tourism has the potential to generate economic benefits for the local community, it can also lead to cultural and environmental degradation if not managed properly. They also describe the different measures that can be taken to promote sustainable tourism practices and to ensure that the festival's cultural authenticity is preserved. Armbrecht, John (2021) noted that cultural festivals have the potential to attract a significant number of tourists to the region, generating economic benefits for the local community. They also describe the different measures that can be taken to promote sustainable tourism practices and to ensure that the economic benefits are distributed equitably. However, they also emphasize the importance of promoting sustainable tourism practices to ensure that the festival's cultural authenticity is preserved and that the economic benefits are distributed equitably. The studies also suggest that tourism can promote socio-economic development in the region, but it must be managed properly to avoid negative impacts on the community and the environment.

Sawian (2012) highlighted the impact of tourism on the social fabric of the indigenous people of Meghalaya, including the Khasi community and the Nongkrem festival. The authors note that while tourism has the potential to generate economic benefits for the local community, it can also lead to cultural and social changes that can have both positive and negative impacts. The Nongkrem festival, for instance, has historically been a celebration of the community's traditional values and practices, and the influx of tourists can either help to promote or dilute those values. Keskin, Sezen, & Kaya (2023) highlighted that tourism has brought about changes in the traditional values, beliefs, and practices of the community and that these changes have had both positive and negative impacts on the community. For instance, while tourism has helped to preserve and promote the Nongkrem festival, it has also led to cultural commodification and commercialization, which can undermine the festival's cultural significance. Keskin, Sezen, & Kaya (2023) pointed out that festivals can have the potential to promote cultural exchange and understanding between different communities, and can help to preserve and promote traditional values and practices. However, they also note that tourism can have negative impacts on the festival's cultural authenticity and can create conflicts between tourists and locals. While tourism can help to promote and preserve the festival, it can also lead to cultural commodification and changes in traditional values and practices. To minimize these negative impacts, the studies suggest promoting sustainable tourism practices that respect and preserve the cultural authenticity of the festival and the community. Additionally, cultural exchange programs can help to promote mutual understanding and respect between tourists and locals.

Som, Suklabaidya, & Chatterjee (2020) highlight the importance of stakeholder engagement in the planning and execution of the festival. They note that stakeholder engagement is crucial for ensuring the cultural authenticity of the festival and promoting its sustainable development. The authors identify various stakeholders, including the Khasi community, government agencies, the tourism industry, and visitors, and emphasize the need for effective communication and collaboration among them. Effective communication and collaboration among stakeholders are crucial for promoting the cultural authenticity and sustainable development of the festival. They emphasize the importance of community engagement and participation in the festival, given its deep roots in the Khasi community.

13.3.1 RESEARCH QUESTION

- What are the challenges faced by the Khasi community in preserving their intangible cultural heritage in the face of modernization and globalization?
- What is the potential of event tourism in promoting the Nongkrem Festival as a tourist destination while ensuring the preservation of its cultural values and traditions?
- What are the guidelines and recommendations for promoting the Nongkrem Festival as an event tourism destination, taking into account the preservation of its cultural values and traditions?

13.3.2 OBJECTIVES

- To identify the challenges faced by the Khasi community in preserving their intangible cultural heritage in the face of modernization and globalization;
- To explore the potential of event tourism in promoting the Nongkrem Festival as a tourist destination while ensuring the preservation of its cultural values and traditions; and
- To develop guidelines and recommendations for promoting the Nongkrem Festival as an event tourism destination.

13.3.3 RESEARCH METHODOLOGY

The aim of this research is to explore how the Nongkrem Dance Festival can be promoted as an event tourism destination while ensuring the preservation of the intangible cultural heritage of the Khasi indigenous community of Meghalaya. The study is using a qualitative research approach to collect data from primary and secondary sources. Primary data is collected through interviews with key stakeholders involved in the organization of the Nongkrem Dance Festival, including the members of the tribe, members of the dance troupe, and representatives from the tourism industry. The interviews will be conducted in person or by telephone with the participants' consent. Secondary data will be collected from various sources such as online databases, government reports, and academic journals. The data collected will provide insights into the cultural heritage of the Khasi community, the history and significance of the Nongkrem Dance Festival, and the potential for event tourism to promote the festival as a tourist destination.

The data collected through telephonic interviews and secondary sources will be analyzed using thematic analysis. The secondary data are analyzed thematically to provide a comprehensive understanding of the research topic and presented in narrative form. Based on the findings of the study, recommendations are developed for promoting the Nongkrem Dance Festival as an event tourism destination while ensuring the preservation of its cultural heritage. The recommendations are developed through a synthesis of the findings from the literature review and interview data.

The research methodology outlined above provides a comprehensive understanding of the potential of event tourism in promoting the Nongkrem Dance Festival as a tourist destination while ensuring the preservation of the intangible cultural heritage of the Khasi indigenous community of Meghalaya. The recommendations developed from the study will provide insights for stakeholders in the tourism industry, policymakers, and the Khasi community on how to promote and preserve the cultural heritage of the Nongkrem Dance Festival.

13.4 THE CONCEPTUAL FRAMEWORK OF INTANGIBLE CULTURAL HERITAGE

The concept of cultural heritage is one of the core elements in the ICH conceptual framework. The cultural practices, customs, and artifacts that a community or group regards as part of their common identity are referred to as cultural heritage. Monuments, structures, crafts, music, dancing, and storytelling are all examples of concrete and intangible aspects (Smith & Akagawa, 2009). As defined by communities, groups, and people as part of their cultural legacy, practices, expressions, knowledge, and skills are referred to as intangible cultural heritage (ICH) (UNESCO, 2003). Communities are considered the main bearers and practitioners of Intangible Cultural Heritage, which emphasizes the need for their active participation in safeguarding and transmitting cultural expressions (Cominelli & Greffe, 2012). Figure 13.1 shows some of the key features related to the understanding of Intangible Cultural Heritage, where recognition and acknowledgment of cultural diversity, and safeguarding and promoting traditional cultural expressions, are very important in order to preserve and conserve the tradition for future generations (Figure 13.1).

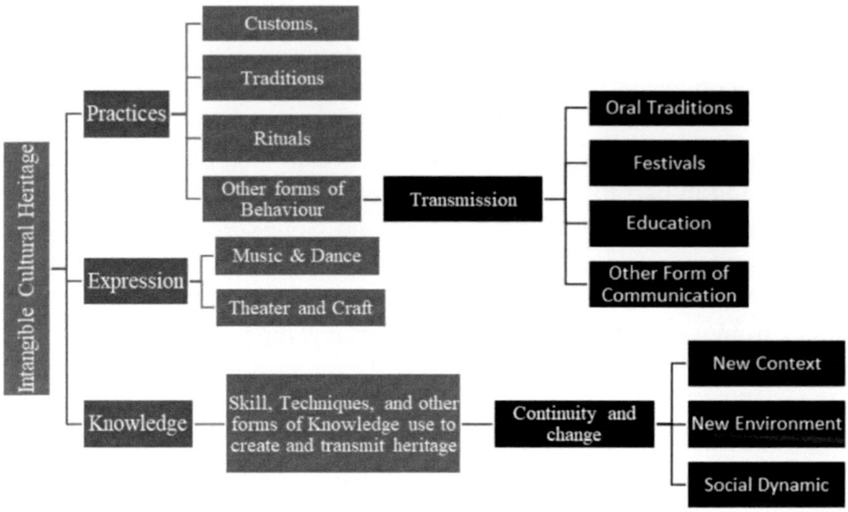

FIGURE 13.1 Intangible cultural heritage.

Sources: Authors.

13.5 CASE STUDY OF NONGKREM FESTIVAL

The Nongkrem Dance Festival is an annual harvest festival celebrated by the Khasi tribe in Meghalaya, India. The festival is held at the Smit village, which is located about 15 km from the capital city of Shillong. It is a significant cultural event that attracts visitors from all over the world. The festival is celebrated in the month of November every year and lasts for four days. This case study aims to explore the history, significance, and cultural importance of the Nongkrem Dance Festival.

13.5.1 HISTORY AND SIGNIFICANCE

The Nongkrem Dance Festival has a long and rich history that dates back several centuries. According to local legends, the festival was started by the ruling Syiem of the Khasi kingdom to appease the gods and goddesses for a good harvest. The festival was initially celebrated by the royal family, but over time, it became a community event that was celebrated by all Khasi people. The festival is also known as Ka Pom-Blang Nongkrem, which means "the dance of the young maidens." It is believed that the festival is a celebration of fertility and prosperity. The festival is also a time for the Khasi people to give thanks to their ancestors and seek their blessings for a prosperous future.

13.5.2 CULTURAL SIGNIFICANCE

The Nongkrem Dance Festival has great cultural significance for the Khasi people. It is a time for them to come together and celebrate their culture, traditions, and customs. The festival is a celebration of the Khasi way of life and serves as a platform for the preservation and promotion of Khasi culture. The festival is also significant for the Khasi community as it helps to maintain their cultural identity in the face of globalization and modernization. The festival is an opportunity for them to showcase their traditional attire, music, and dance to the world.

13.5.3 CELEBRATION

The Nongkrem Dance Festival is celebrated over four days. The festival begins with the sacrifice of a goat at the sacred altar of the Syiem, who is the head of the Khasi tribe. The goat is sacrificed to appease the gods and goddesses and seek their blessings for a good harvest.

The second day of the festival is known as Ka Shad Kynthei, which means "the day of the young maidens." On this day, the unmarried young women of the community perform a dance known as the Ka Shad Kynthei dance. The dance is performed to celebrate their youth and beauty.

The third day of the festival is known as Ka Shad Suk Mynsiem, which means "the day of the men." On this day, the men of the community perform a dance known as the Ka Shad Suk Mynsiem dance. The dance is performed to celebrate the harvest and the prosperity of the community.

The fourth and final day of the festival is known as Ka-Shad-Phur-Kan, which means "the day of the grand procession." On this day, the entire community comes together to participate in a grand procession. The procession is led by Syiem and his entourage, who are accompanied by the young maidens and the men of the community. The procession is accompanied by traditional music and dance and ends at the sacred altar of the Syiem.

13.5.4 INITIATIVES AND POLICY IMPLEMENTATION FOR THE PRESERVATION OF INTANGIBLE HERITAGE MEGHALAYA

Meghalaya has taken several initiatives and policy implementations for the preservation of its intangible cultural heritage. Some of the key initiatives and policies are:

- **Meghalaya State Policy on Intangible Cultural Heritage (2015):** The state government of Meghalaya has developed a policy framework to safeguard and promote the state's intangible cultural heritage. The policy emphasizes the need to document and preserve traditional knowledge, customs, and practices, and to promote them as a means of fostering social cohesion and economic development, including the Meghalaya Ancient Monuments and Records Act, which aims to protect and preserve the state's ancient monuments and records.

- **Meghalaya State Council for Cultural Affairs (MSCCA):** The MSCCA was established in 1973 to promote and preserve the cultural heritage of Meghalaya, which includes the intangible cultural heritage.
- **Directorate of Arts and Culture:** The Directorate of Arts and Culture was established in Meghalaya in 1976 to promote and preserve the state's rich cultural heritage. The Directorate plays a key role in organizing cultural events and festivals, supporting traditional artists and artisans, and promoting cultural tourism.
- **Promotion of Community-Based Tourism:** Meghalaya has recognized the potential of community-based tourism as a means of promoting the state's cultural heritage while also providing economic opportunities to local communities. The state government has supported the development of community-based tourism initiatives, such as homestays, village tourism, and cultural tours.
- **Documentation and Digitization of Intangible Cultural Heritage:** Meghalaya has undertaken several initiatives to document and digitize its intangible cultural heritage. The state government has supported the development of archives, museums, and cultural centers that showcase the state's traditional practices and customs.
- **Support for Traditional Artisans and Craftspeople:** Meghalaya has a rich tradition of handicrafts and traditional arts, such as weaving, bamboo crafts, and woodcarving. The state government has provided support to traditional artisans and craftspeople by organizing training programs, providing marketing opportunities, and promoting their products as part of cultural tourism.
- **Development of Cultural Infrastructure:** Meghalaya has invested in the development of cultural infrastructure, such as cultural centers, museums, and performance spaces, to promote and preserve its intangible cultural heritage.

- **Inclusion of ICH in School Curriculum:** Meghalaya's government has included ICH in the school curriculum to promote awareness and appreciation of the state's cultural heritage among the younger generation.
- **Promotion of ICH Through Tourism:** Meghalaya's government has promoted the state's ICH as a tourist attraction, including the Nongkrem dance festival, Behdiengkhlam, and Wangala Dance which are the major tourist attractions that showcase the state's traditional cultural practices.

Overall, Meghalaya's initiatives and policy implementations for the preservation of its intangible cultural heritage demonstrate a commitment to safeguarding and promoting the state's rich cultural traditions and practices. These initiatives have the potential to not only preserve cultural heritage but also to promote sustainable development and foster social cohesion.

13.5.5 CHALLENGES OF THE NONGKREM FESTIVAL

The Nongkrem festival is a significant cultural event celebrated by the Khasi tribe in Meghalaya, India. It is a harvest festival that is held every year, usually in November, and involves various rituals and ceremonies. This literature review aims to explore the major challenges associated with the Nongkrem festival. One of the major challenges of the Nongkrem festival is the declining participation of young people in the celebrations. According to Rizvi & Roy (2006) Khasi tribe of Meghalaya. B.R. Publishing Corporation (2018), there is a growing concern among the Khasi community that younger generations are becoming disconnected from their cultural heritage and are not interested in participating in the traditional ceremonies associated with the festival. This could threaten the sustainability of the festival and the preservation of the Khasi culture. Lyngdoh (2022) also stated that the festival has undergone several changes over the years, and there is a risk that it may lose its cultural significance if it is not celebrated in its original form. The author recommends that efforts should be made to preserve the original rituals and traditions associated with the festival.

Another challenge is the impact of modernization and tourism on the festival. The Nongkrem festival has become an important tourist attraction in recent years, and there is a concern that the increased commercialization and commodification of the festival could have negative impacts on its cultural significance. In a study by Diengdoh (2016), it was found that the festival had become a platform for commercial activities such as the sale of souvenirs, which could dilute the traditional aspects of the festival and compromise its cultural authenticity. The authors argue that this has led to a loss of cultural authenticity and recommend that efforts should be made to promote the cultural significance of the festival.

Additionally, the Nongkrem festival faces challenges related to the management and organization of the event. As the festival has become more popular, there is a need for better coordination and planning among the various stakeholders involved in the organization of the event. In a study by Lonardi & Unterpertinger (2022), it was found that the lack of coordination among different departments, such as tourism, cultural affairs, and local government, has resulted in inefficiencies and conflicts in the organization of the festival and led to the failure in crowd handling. Since the festival attracts a large number of visitors, there is a risk of overcrowding and stampedes. In a study by Lyngdoh (2022), it was found that the lack of proper infrastructure and crowd control measures was a major concern during the festival. The authors recommend that measures such as the installation of CCTV cameras and the deployment of additional security personnel should be taken to ensure the safety of visitors.

The issue of waste management is also a significant challenge during the Nongkrem festival. The festival generates a significant amount of waste, including plastic bags, food waste, and other forms of litter. In a study by Lyngdoh (2022), it was found that the lack of proper waste management facilities was a major concern during the festival. The authors recommend that measures such as the installation of waste bins and the promotion of eco-friendly practices should be taken to address this issue.

Furthermore, the issue of commercialization is another challenge associated with the Nongkrem festival. According to Lyngdoh (2022), the festival has become increasingly commercialized, with vendors selling souvenirs, food, and other items. The authors argue that this has led to a loss of cultural authenticity and recommend that efforts should be made to promote the cultural significance of the festival.

Finally, the Nongkrem festival faces challenges related to the preservation and transmission of traditional knowledge and skills. The festival involves various traditional practices, such as dance performances, traditional dress, and music, which are passed down from generation to generation. However, there is a concern that the younger generations may not be interested in learning and practicing these traditional skills, which could result in the loss of cultural heritage. In a study by Diengdoh (2016), it was found that there is a need to promote traditional knowledge and skills among younger generations and to ensure that they are given opportunities to participate in the festival. Addressing these challenges will be crucial in ensuring the sustainability and preservation of this important cultural event.

13.6 HERITAGE PROMOTION THROUGH EVENT TOURISM DESTINATION PERSPECTIVE

The event tourism destination perspective is a strategy for promoting heritage through tourism, where cultural and historical events are organized to attract tourists and promote a destination's heritage. This approach recognizes that cultural events can be powerful tools for promoting a destination's heritage and can contribute to the development of the tourism industry. Heritage promotion through event tourism destination perspective is a rapidly growing area of research. Event tourism is a subset of tourism that involves traveling to attend events, festivals, and other activities. Heritage promotion, on the other hand, is the practice of preserving and promoting cultural heritage sites and traditions. In this literature review, we will explore various studies and research papers that focus on the topic of heritage promotion from an event tourism destination perspective.

One of the most important aspects of heritage promotion through event tourism destination perspective is the economic impact of events on local communities. According to a study by Richards & Wilson (2007), events can bring significant economic benefits to local communities, including increased spending, employment opportunities, and tax revenues. The authors argue that the economic impact of events can be maximized by promoting local heritage and cultural assets, which can attract a larger number of tourists. Another study by Gursoy & Kendall (2006) examined the role of cultural events in promoting heritage tourism. The authors found that cultural events can be effective in attracting tourists to heritage sites, but that the success of such events depends on factors such as the quality of the event, the attractiveness of the destination, and the marketing efforts of local authorities. In addition to economic benefits, heritage promotion through event tourism destination perspective can also have social and cultural benefits. A study by Yeoman et al. (2012) explored the cultural and social impact of festivals on local communities. The authors found that festivals can provide opportunities for social interaction and cultural exchange, as well as promote cultural heritage. However, the success of heritage promotion through the event tourism destination perspective also depends on effective planning and management. According to a study by Getz and Page (2016), effective planning and management are essential for ensuring the sustainability of events and minimizing negative impacts on local communities and the environment. The authors argue that effective planning and management should include stakeholder consultation, environmental impact assessments, and risk management strategies.

Overall, heritage promotion through an event tourism destination perspective is an important area of research that has significant economic, social, and cultural implications. The success of this approach depends on effective planning, management, and marketing strategies that promote local heritage and cultural assets. Several studies have highlighted the importance of heritage promotion through the event tourism destination perspective. According to Getz and Page (2016), heritage events can provide a platform for the preservation and promotion of cultural heritage, as well as for the development of the local economy. Furthermore, the events can contribute to the creation of a sense of community and pride among residents, as well as provide an opportunity for visitors to learn about local culture and history.

13.7 PRESERVING INTANGIBLE CULTURAL HERITAGE

It is a crucial aspect of ensuring that the diverse cultural traditions, practices, and knowledge of communities around the world are passed down to future generations. Intangible cultural heritage includes practices, customs, knowledge, and skills that are transmitted from one generation to the next, often through oral traditions, performance, and social interaction.

There are several ways to preserve intangible cultural heritage, including:

- **Documentation:** Recording and documenting intangible cultural heritage through various media such as photographs, videos, audio recordings, and written materials is an essential step in preserving it for future generations. This documentation can provide valuable information for researchers, scholars, and communities.
- **Transmission:** Ensuring that intangible cultural heritage is transmitted from one generation to the next is critical for its survival. This can be done through various means, including traditional apprenticeships, workshops, and community events that encourage participation and knowledge sharing.
- **Education:** Introducing intangible cultural heritage into formal education systems can help to ensure that future generations learn about and appreciate the importance of these cultural practices.

- **Cultural Revitalization:** Revitalizing cultural practices that have been lost or are in danger of disappearing can be achieved through various means, including community-based initiatives, festivals, and events that celebrate and promote intangible cultural heritage.
- **Legal Protection:** Providing legal protection to intangible cultural heritage through various means such as UNESCO's as highlighted in the Convention for the Safeguarding of the Intangible Cultural Heritage, national laws, and policies can help to ensure its survival.

Overall, preserving intangible cultural heritage is essential for maintaining cultural diversity, promoting understanding between different cultures, and enriching the lives of people around the world.

13.7.1 THE 6 H'S OF CULTURAL HERITAGE

The 6 H's of cultural heritage are a framework used to understand the different dimensions of heritage. The 6 H's stand for:

- **Historicity:** This refers to the historical significance of heritage, including the events, people, and places that have shaped the past. Historicity helps us understand the context and meaning of cultural heritage.
- **Humanity:** Cultural heritage is created and maintained by people, and reflects the values, beliefs, and practices of different communities. The human dimension of heritage recognizes the importance of community participation in heritage preservation.
- **Holism:** Cultural heritage is complex and interconnected, encompassing a range of tangible and intangible elements. Holism emphasizes the need to view heritage in a holistic manner, recognizing the interdependence of different elements.
- **Heritage Value:** Cultural heritage has different values, including historical, social, cultural, and economic. Understanding the different values of heritage can help us prioritize and manage heritage resources.
- **Heritage Management:** Managing cultural heritage involves a range of activities, including documentation, conservation, interpretation, and dissemination. Effective heritage management requires collaboration and participation from a range of stakeholders.

- **Heritage Ethics:** Cultural heritage is not neutral and can have different meanings and interpretations for different communities. Heritage ethics emphasizes the need to approach heritage in a respectful and responsible manner, taking into account the perspectives and interests of all stakeholders (Figure 13.2).

FIGURE 13.2 6 H's cultural heritage.

Sources: Author.

Overall, the 6 H's of cultural heritage provide a useful framework for understanding the different dimensions of heritage and can help guide heritage management and preservation efforts. Heritage promotion through event tourism has gained immense popularity in recent years, as it is considered an effective way to promote tourism and preserve cultural heritage. This literature review aims to explore the different perspectives related to the promotion of heritage through event tourism destinations. Event tourism is an important component of the tourism industry, and it is often used as a tool for promoting and developing tourism destinations. Heritage tourism, on the other hand, is a form of tourism that focuses on the history, culture, and traditions of a particular place. Heritage promotion through event tourism destinations perspective combines these two concepts to promote heritage sites, buildings, monuments, and cultural traditions through various events. In addition, several studies have explored the impact of heritage events on the local economy. For example, a study by Naehyun, Hyuckgi & Sangmo (2013) found that heritage events can have a significant impact on local businesses, as they attract a large number of tourists and generate revenue for local businesses such as hotels, restaurants, and shops.

However, heritage promotion through an event tourism destination perspective is not without its challenges. One of the main challenges is the need to balance tourism development with the preservation of cultural heritage. According to Timothy (2011), tourism development can sometimes lead to the commodification of culture, which can have negative impacts on the authenticity and integrity of cultural heritage sites and traditions.

Another challenge is the need to ensure that heritage events are sustainable and do not have negative impacts on the environment and local communities. In a study by Nocca (2017), it was found that the sustainability of heritage events can be enhanced through the use of eco-friendly practices, such as reducing waste and promoting energy efficiency.

In conclusion, heritage promotion through event tourism destination perspective has emerged as an effective way to promote and preserve cultural heritage. However, it is important to ensure that heritage events are sustainable, inclusive, and respectful of local communities and their cultural traditions.

13.7.2 FINDING

The festival allows the Khasi community to showcase their cultural practices, such as traditional music, dance, attire, and rituals. At the festival, one will also experience the traditions that have been passed on from generation to generation without any changes or alterations in the rituals done by the King (Syiem) and his ministers under the purview of the Queen Mother (syiemsad).

According to the study, the festival has the potential to contribute to the growth of event tourism in the region. The study also found that the Khasi Indigenous community can use the festival to promote their Intangible cultural heritage, which attracts a lot of tourists and generates revenue for the local economy. However, the study notes that there are challenges that must be addressed in order to ensure the festival's sustainability and the preservation of Khasi cultural heritage.

The following are some of the key findings from this study:

1. **Loss of Traditional Knowledge:** The Khasi community is facing a loss of traditional knowledge due to the shift towards modern education systems and lifestyle changes. The younger generations are less interested in learning about their traditional practices, and there is a lack of formal documentation of these practices.

2. **Influence of Modernization and Globalization:** The influence of modernization and globalization has resulted in the erosion of traditional practices, including language, music, dance, and art. The Khasi community is exposed to external cultures and values that can cause them to overlook their cultural heritage.

3. **Economic Development and the Benefits:** The rapid economic development in the region has resulted in the loss of land and natural resources, which are integral to the Khasi culture. The exploitation of these resources has led to the destruction of sacred sites, forests, and water bodies that are significant to the Khasi community. However, event tourism can provide significant economic benefits to the local community, including job opportunities and income generation. Tourists attending the festival can also contribute to the local economy by spending money on accommodations, food, souvenirs, and other local products and services.

4. **Limited Institutional Support:** There is limited institutional support and funding for the preservation of the Khasi intangible cultural heritage. There is a lack of awareness and understanding of the importance of cultural heritage preservation among policymakers and the general public.

5. **Intergenerational Communication Gap:** There is a communication gap between the older and younger generations, resulting in the loss of traditional practices and knowledge. The younger generations are not fully aware of the significance of their cultural heritage and are not actively involved in preserving it.

6. **Unique Cultural Experience:** The Nongkrem Festival offers a unique cultural experience to tourists, with its traditional rituals, music, dance, and costumes. It is an opportunity to learn about the Khasi culture and traditions, which can attract tourists interested in cultural tourism.

7. **Infrastructure Development:** The promotion of the Nongkrem Festival as a tourist destination can also lead to the development of necessary infrastructure such as roads, transportation, accommodations, and tourist facilities. This can benefit both tourists and the local community.
8. **Preserving Cultural Heritage and Authenticity:** While promoting the Nongkrem Festival as a tourist destination, it is essential to ensure the preservation of its cultural values and traditions such as traditional rituals, music, dance, and costumes that make the festival unique. This can be achieved through community involvement in the planning and management of the festival and also by considering decision-making processes by promoting responsible tourism practices.
9. **Challenges in Promoting Event Tourism:** There are challenges in promoting event tourism, such as ensuring the safety and security of tourists, managing the influx of visitors, and preserving the environment and natural resources. These challenges need to be addressed to ensure the sustainability of event tourism.
10. **Leverage Technology and Social Media:** Technology and social media can be used to promote the Nongkrem Festival as an event tourism destination. Social media platforms can be used to create buzz and awareness about the festival, while technology can be used to provide information and facilitate bookings.

These guidelines and recommendations can help in promoting the Nongkrem Festival as an event tourism destination while ensuring the preservation of its cultural values and traditions. By focusing on cultural authenticity, community involvement, responsible tourism, necessary infrastructure, safety and security, and leveraging technology and social media, the Nongkrem Festival can attract more tourists and contribute to the local economy. This approach can also ensure that the festival benefits the Khasi community, respects their cultural practices and traditions, and promotes responsible tourism.

13.8 CONCLUSION

Nongkrem Dance Festival plays a significant role in preserving the intangible cultural heritage of the Khasi indigenous community in Meghalaya.

The festival provides an opportunity for the community to showcase their cultural practices, including traditional music, dance, attire, and rituals. Furthermore, the festival has the potential to contribute to the development of event tourism in the region and generate income for the local economy.

However, as highlighted above, in order to ensure the festival's continued existence and the authenticity of all the rituals and traditions of this festival, which is considered the essential intangible cultural heritage of the Khasi community, a collaborative approach involving the Khasi community, government, and tourism stakeholders is required. This approach can ensure the preservation of the festival, respect the cultural practices and traditions, and promote responsible tourism.

KEYWORDS

- **cultural practices**
- **intangible cultural heritage**
- **Khasi Indigenous Community**
- **Meghalaya State Council for Cultural Affairs**
- **Nongkrem dance**
- **tourism destination**
- **tourism industry**
- **tourism stakeholders**

REFERENCES

Armbrecht, J. (2021). Event quality, perceived value, satisfaction and behavioral intentions in an event context. *Scandinavian Journal of Hospitality and Tourism, 21*(2), 169–191. https://doi.org/10.1080/15022250.2021.1877191.

Basaiamoit, A., & Khyriemmujat, S. (2016). The Pomblang: A Depiction of Khasi Polity. In *Revisiting Traditional Institutions in the Khadi-Jaintia Hills* (2016th ed., pp. 38–51). Cambridge Scholars.

Cominelli, F., & Greffe, X. (2012). Intangible cultural heritage: Safeguarding for creativity. *City, Culture and Society, 3*(4), 245–250.

Diengdoh, N. J. A. (2016). Attributions through Performing Arts in the Khasi Harvest Dance. *Malaysian Journal of Performing and Visual Arts, 2*(1), 25–42.

Dutta, M. (2019). Majuli: An emerging niche tourism destination village in North East India. In U. N. Shukla & S. K. Kulshreshtha (Eds.), *Emerging Dynamics of Indian*

Tourism & Hospitality: Transformation & Innovation. Gaziabad: Copal Publishing. ISBN 9789383419562.

Dutta, P., & Ghosh, J. (2016). (Re) Negotiating Identities through Festivals: An Account of Khasis and Konyaks. *Anthropology Today: An International Peer Reviewed NEIRA Journal, 2454–2709*, 1, 41–59.

Dychkovskyy, S., & Ivanov, S. (2020). Festival Tourism as Part of International Tourism and a Factor in the Development of Cultural Tourism. *Informacijos mokslai, 89*, 73–82. https://doi.org/10.15388/Im.2020.89.41.

Getz, D. (2008). Event tourism: Definition, evolution, and research. *Tourism Management, 29*(3), 403–428. https://doi.org/10.1016/j.tourman.2007.07.017.

Getz, D., & Page, S. J. (2016). Progress and Prospects for Event Tourism Research. *Tourism Management, 52*, 593–631.

Gursoy, D., & Kendall, K. W. (2006). Hosting Mega-Events: Modeling Locals' Support. *Journal of Travel Research, 45*(2), 283–294. https://doi.org/10.1177/0047287506286753.

Hafstein, V. T. (2012). Cultural Heritage. In *A Companion to Folklore* (pp. 500–519). John Wiley and Sons Inc. https://doi.org/10.1002/9781118379936.ch26.

Isar, Y. R. (2006). Cultural diversity. *Theory, Culture & Society, 23*(2–3), 372–375.

Keskin, E., Sezen, N., & Kaya, Ö. (2023). The Role of Festivals in Destination Branding. In *Advances in Hospitality, Tourism, and the Services Industry* (pp. 1–17). IGI Global. https://doi.org/10.4018/978-1-6684-6356-7.ch001.

Liu, Y. (2022). Application of Digital Technology in Intangible Cultural Heritage Protection. *Mobile Information Systems, 2022*, 1–8. https://doi.org/10.1155/2022/7471121.

Lonardi, S., & Unterpertinger, Y. (2022). The Relevance of Intangible Cultural Heritage and Traditional Languages for the Tourism Experience: The Case of Ladin in South Tyrol. *Sustainability, 14*(5), 2729. https://doi.org/10.3390/su14052729.

Lyngdoh, B. (2022). An Exploration of Tourism Development of Meghalaya, India: What is the way forward? *Atna Journal of Tourism Studies, 17*(1), 143–166.

Lyngdoh, M. P. R. (2013). The Rong Khil (Tiger Festival) of the War Jaintias. Retrieved from http://indpaedia.com/ind/index.php/The_Rong_Khil_(Tiger_Festival)_of_the_War_Jaintias#The_Origin_of_the_Festival on Retrieved on 8/6/2024.

Mair, J., & Duffy, M. (2014). Social cohesion and local festivals. *Journal of Policy Research in Tourism, Leisure & Events, 7*(3), 282–298.

Mon S., & Thanikkad, J. (2021). Cultural tourism products in Kerala – An exploratory study of tourist's satisfaction level and loyalty towards cultural tourism in Kerala. Retrieved from: https://www.researchgate.net/publication/338450001 (accessed on 18 June 2024).

Morales, F. C. (2020). Tourism: An important sector for the World Economy – Global Tourism Forum. *Global Tourism Forum*. Retrieved from: https://www.globaltourismforum.org/blog/2020/02/06/tourism-an-important-sector-for-the-world-economy/.

Naehyun, J., Lee, H., & Lee, S. (2013). Event Quality, Perceived Value, Destination Image, and Behavioral Intention of Sports Events: The Case of the IAAF World Championship, Daegu, 2011. *Asia Pacific Journal of Tourism Research, 18*(8), 849–864. https://doi.org/10.1080/10941665.2012.711336.

Nayak, P., & Mishra, S. K. (2013, March). Problems and prospects of promoting tourism in Meghalaya. In *National Seminar On 'Promotion of International Tourism Circuits in North East India: Prospects, Priorities and Strategic Options' Organized by The Department of Business Administration, Assam University, Silchar, India* (pp. 8–9).

Nocca, F. (2017). The Role of Cultural Heritage in Sustainable Development: Multidimensional Indicators as Decision-Making Tool. *Sustainability, 9*(10), 1882. https://doi.org/10.3390/su9101882.

Richards, G., & Wilson, J. (2007). Tourism Development Trajectories: From Culture to Creativity? In G. Richards & J. Wilson (Eds.), *Tourism, Creativity and Development* (pp. 1–34). https://doi.org/10.1787/9789264040731-en.

Rizvi, S. H. M., & Roy, S. (2006). *Khasi tribe of Meghalaya.* B.R. Publishing Corporation. ISBN: 9788176465229.

Sawian, S. S. (2012). Reflection of polity in the nongkrem dance. *The Dorbar of Hima Khyrim Smit Souvenir*.

Sinha, R., & Murdia, M. (2018). Heritage Festivals: An important factor to promote cultural tourism (A study of Murshidabad, West Bengal). *Annals of Art, Culture & Humanities, 3*(2), 9–1. Online ISSN-2455-5843.

Smith, L., & Akagawa, N. (Eds.). (2008). *Intangible Heritage* (1st ed.). Routledge. https://doi.org/10.4324/9780203884973.

Som, J., Suklabaidya, P., & Chatterjee, S. (2020). Stakeholders' Perspective on Tourism Infrastructure at Khajuraho Dance Festival. *Global Journal of Enterprise Information System, 12*, 82–90. https://doi.org/10.18311/gjeis/2020.

Thanikkad, J., & Saleem, S. (2021). Cultural Tourism Products in Kerala: An exploratory Study of Tourist's Satisfaction Level and Loyalty towards Cultural Tourism in Kerala. *International Journal of Tourism and Hospitality, 1*(4), 1–10.

Timothy, D. (2011). *Cultural Heritage and Tourism: An Introduction.* Channel View Publications. ISBN: 9781845411787, https://doi.org/10.21832/9781845411787.

Tynsong, I. (2019). Connecting cultural and festival in preserving the identity: A case of Nongkrem festival. *Journal of the Gujarat Research Society, 21*(10), 965–975.

UNESCO. (2014). Convention for the Safeguarding of the Intangible Cultural Heritage. In *Basic texts: Of the 2003 Convention for the Safeguarding of the Intangible Cultural Heritage*. Essay.

Yeoman, I., Robertson, M., Ali-Knight, J., Drummond, S., & McMahon-Beattie, U. (Eds.). (2003). *Festival and Events Management* (1st ed.). Routledge. https://doi.org/10.4324/9780080477701.

INDEX

C

D

N

O

P

Q

T

V

W

X

Y

Z